Environanotechnology

Environanotechnology

Dr. Siya Ram

RANDOM

RANDOM PUBLICATIONS

NEW DELHI (INDIA)

Environanotechnology

ISBN 978-93-5111-954-8

Published in 2016 in India by

RANDOM PUBLICATIONS

4376-A/4B, Gali Murari Lal, Ansari Road
New Delhi-110 002
Phone : +9111-43580356, 011-23289044, 011-43142548
e-mail: sales@randompublications.com,
info@randompublications.com, randomexports@gmail.com

Reprinted 2024

Type Setting by : Friends Media, Delhi-110089
Digitally Printed at : Replika Press Pvt. Ltd.

Preface

The environmental impact of nanotechnology is the possible effects that the use of nanotechnological materials and devices will have on the environment. As nanotechnology is an emerging field, there is great debate regarding to what extent industrial and commercial use of nanomaterials will affect organisms and ecosystems. Nanotechnology's environmental impact can be split into two aspects: the potential for nanotechnological innovations to help improve the environment, and the possibly novel type of pollution that nanotechnological materials might cause if released into the environment.

Nanotechnological products, processes and applications are expected to contribute significantly to environmental and climate protection by saving raw materials, energy and water as well as by reducing greenhouse gases and hazardous wastes. Using nanomaterials therefore promises certain environmental benefits and sustainability effects. Note, however, that nanotechnology currently plays a rather subordinate role in environmental protection, whether it be in research or in practical applications. Environmental engineering companies themselves attach only limited importance to nanotechnology in their respective fields.

The aim of this book is to report on the results recently achieved in different countries. It provides useful technological information for environmental scientists and will assist them in creating cost-effective nanotechnologies to solve critical environmental problems, including those associated with energy production.

– *Author*

Contents

1

Introduction

Nanotechnology is defined as fabrication of devices with atomic or molecular scale precision. Devices with minimum feature sizes less than 100 nanometers (nm) are considered to be products of nanotechnology. A nanometer is one billionth of a meter (10 – 9 m) and is the unit of length that is generally most appropriate for describing the size of single molecules. The nanoscale marks the nebulous boundary between the classical and quantum mechanical worlds; thus, realization of nanotechnology promises to bring revolutionary capabilities.

Fabrication of nanomachines, nanoelectronics and other nanodevices will undoubtedly solve an enormous amount of the problems faced by mankind today. Nanotechnology is currently in a very infantile stage. However, we now have the ability to organize matter on the atomic scale and there are already numerous products available as a direct result of our rapidly increasing ability to fabricate and characterize feature sizes less than 100 nm. Mirrors that don't fog, biomimetic paint with a contact angle near 180°, gene chips and fat soluble vitamins in aqueous beverages are some of the first manifestations of nanotechnology. However, immenant breakthroughs in computer science and medicine will be where the real potential of nanotechnology will first be achieved. Nanoscience is an interdisciplinary field that seeks to bring about mature nanotechnology.

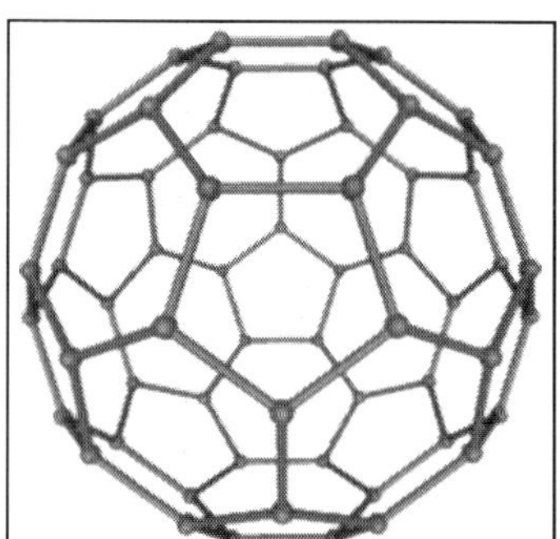

Fig. Buckminsterfullerene

Focusing on the nanoscale intersection of fields such as physics, biology, engineering, chemistry, computer science and more, nanoscience is rapidly expanding. Nanotechnology centers are popping up around the world as more funding is provided and nanotechnology market share increases.

Nanotechnology is extremely diverse, ranging from novel extensions of conventional device physics, to completely new approaches based upon molecular self-assembly, to developing new materials with dimensions on the nanoscale, even to speculation on whether we can directly control matter on the atomic scale.

There has been much debate on the future of implications of nanotechnology. Nanotechnology has the potential to create many new materials and devices with wide-ranging applications, such as in medicine, electronics, and energy production. On the other hand, nanotechnology raises many of the same issues as with any introduction of new technology, including concerns about the toxicity and environmental impact of nanomaterials and their potential effects on global economics, as well as speculation about various doomsday scenarios.

These concerns have led to a debate among advocacy groups and governments on whether special regulation of nanotechnology is warranted. Feynman described a process by which the ability to manipulate individual atoms and molecules might be developed, using one set of precise tools to build and operate another proportionally smaller set, so on down to the needed scale.

In the course of this, he noted, scaling issues would arise from the changing magnitude of various physical phenomena: gravity would become less important, surface tension and Van der Waals attraction would become more important, etc.

FUNDAMENTAL CONCEPTS

One nanometer (nm) is one billionth, or 10^{-9}, of a meter. By comparison, typical carbon-carbon bond lengths, or the spacing between these atoms in a molecule, are in the range 0.12-0.15 nm, and a DNA double-helix has a diameter around 2 nm. On the other hand, the smallest cellular lifeforms, the bacteria of the genus Mycoplasma, are around 200 nm in length.

To put that scale in another context, the comparative size of a nanometer to a meter is the same as that of a marble to the size of the earth. Or another way of putting it: a nanometer is the amount a man's beard grows in the time it takes him to raise the razor to his face.

Two main approaches are used in nanotechnology. In the "bottom-up" approach, materials and devices are built from molecular components which assemble themselves chemically by principles of molecular recognition. In the "top-down" approach, nano-objects are constructed from larger entities without atomic-level control.

LARGER TO SMALLER: A MATERIALS PERSPECTIVE

A number of physical phenomena become pronounced as the size of the system decreases. These include statistical mechanical effects, as well as quantum mechanical effects, for example the "quantum size effect" where the electronic properties of solids are altered with great reductions in particle size. This effect does not come into play by going from macro to micro dimensions. However, it becomes dominant when the nanometer size range is reached.

Additionally, a number of physical (mechanical, electrical, optical, etc.) properties change when compared to macroscopic systems. One example is the increase in surface area to volume ratio altering mechanical, thermal and catalytic properties of materials. Novel *mechanical* properties of nanosystems are of interest in the nanomechanics research. The catalytic activity of nanomaterials also opens potential risks in their interaction with biomaterials.

Materials reduced to the nanoscale can show different properties compared to what they exhibitf on a macroscale, enabling unique applications. For instance, opaque substances become transparent (copper); stable materials turn combustible (aluminum); solids turn into liquids at room temperature (gold); insulators become conductors (silicon).

A material such as gold, which is chemically inert at normal scales, can serve as a potent chemical catalyst at nanoscales. Much of the fascination with nanotechnology stems from these quantum and surface phenomena that matter exhibits at the nanoscale.

Simple to Complex: a Molecular Perspective

Modern synthetic chemistry has reached the point where it is possible to prepare small molecules to almost any structure. These methods are used today to produce a wide variety of useful chemicals such as pharmaceuticals or commercial polymers.

This ability raises the question of extending this kind of control to the next-larger level, seeking methods to assemble these single molecules into supramolecular assemblies consisting of many molecules arranged in a well defined manner. These approaches utilize the concepts of molecular self-assembly and/or supramolecular chemistry to automatically arrange themselves into some useful conformation through a bottom-up approach. The concept of molecular recognition is especially important: molecules can be designed so that a specific conformation or arrangement is favoured due to non-covalent intermolecular forces.

The Watson-Crick basepairing rules are a direct result of this, as is the specificity of an enzyme being targeted to a single substrate, or the specific folding of the protein itself. Thus, two or more components can be designed to be complementary and mutually attractive so that they make a more complex and useful whole. Such bottom-up approaches should be able to produce devices in parallel and much cheaper than top-down methods, but could potentially be overwhelmed as the size and complexity of the desired assembly increases.

Most useful structures require complex and thermody-namically unlikely arrangements of atoms. Nevertheless, there are many examples of self-assembly based on molecular recognition in biology, most notably Watson-Crick basepairing and enzyme-substrate interactions. The challenge for nanotechnology is whether these principles can be used to engineer novel constructs in addition to natural ones.

Molecular Nanotechnology: a Long-term View

Molecular nanotechnology, sometimes called molecular manufacturing, is a term given to the concept of engineered nanosystems (nanoscale machines) operating on the molecular scale. It is especially associated with the concept of a molecular assembler, a machine that can produce a desired structure or device atom-by-atom using the principles of mechanosynthesis.

Manufacturing in the context of productive nanosystems is not related to, and should be clearly distinguished from, the conventional technologies used to manufacture nanomaterials such as carbon nanotubes and nanoparticles.

When the term "nanotechnology" was independently coined and popularized by Eric Drexler (who at the time was unaware of an earlier usage by Norio Taniguchi) it referred to a future manufacturing technology based on molecular machine systems.

The premise was that molecular scale biological analogies of traditional machine components demonstrated molecular machines were possible: by the countless examples found in biology, it is known that sophisticated, stochastically optimised biological machines can be produced..

It is hoped that developments in nanotechnology will make possible their construction by some other means, perhaps using biomimetic principles. However, Drexler and other researchers have proposed that advanced nanotechnology, although perhaps initially implemented by biomimetic means, ultimately could be based on mechanical engineering principles, namely, a manufacturing technology based on the mechanical functionality of these components (such as gears, bearings, motors, and structural members) that would enable programmable, positional assembly to atomic specification (PNAS-1981).

The physics and engineering performance of exemplar designs were analyzed in Drexler's book *Nanosystems*. In general it is very difficult to assemble devices on the atomic scale, as all one has to position atoms are other atoms of comparable size and stickyness.

Another view, put forth by Carlo Montemagno, is that future nanosystems will be hybrids of silicon technology and biological molecular machines. Yet another view, put forward by the late Richard Smalley, is that mechanosynthesis is impossible due to the difficulties in mechanically manipulating individual molecules.

TOOLS AND TECHNIQUES IN NANOTECHNOLOGY

A microfabricated cantilever with a sharp tip is deflected by features on a sample surface, much like in a phonograph but on a much smaller scale. A laser beam reflects off the backside of the cantilever into a set of photodetectors, allowing the deflection to be measured and assembled into an image of the surface.The first observations and size measurements of nano-particles were made during the first decade of the 20th century.

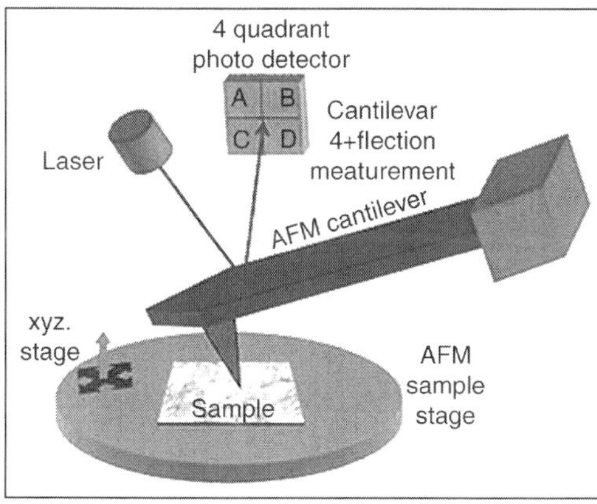

Fig. Typical AFM Setup.

They are mostly associated with the name of Zsigmondy who made detailed studies of gold sols and other nanomaterials with sizes down to 10 nm and less. He published a book in 1914.[23] He used ultramicroscope that employs a *dark field* method for seeing particles with sizes much less than light wavelength.

There are traditional techniques developed during 20th century in Interface and Colloid Science for characterizing nanomaterials. These are widely used for *first generation* passive nanomaterials specified in the next section.

These methods include several different techniques for characterizing particle size distribution. This characterization is imperative because many materials that are expected to be nano-sized are actually aggregated in solutions. Some of methods are based on light scattering.

Other apply ultrasound, such as ultrasound attenuation spectroscopy for testing concentrated nano-dispersions and microemulsions.[24] There is also a group of traditional techniques for characterizing surface charge or zeta potential of nano-particles in solutions.

Next group of nanotechnological techniques include those used for fabrication of nanowires, those used in semiconductor fabrication such as deep ultraviolet lithography, electron beam lithography, focused ion beam machining, nanoimprint lithography, atomic layer deposition, and molecular vapour deposition, and further including molecular self-assembly techniques such as those employing di-block copolymers. However, all of these techniques preceded the nanotech era, and are extensions in the development of scientific advancements rather than techniques which were devised with the sole purpose of creating nanotechnology and which were results of nanotechnology research.

There are several important modern developments. The atomic force microscope (AFM) and the Scanning Tunneling Microscope (STM) are two early versions of scanning probes that launched nanotechnology.

There are other types of scanning probe microscopy, all flowing from the ideas of the scanning confocal microscope developed by Marvin Minsky in 1961 and the scanning acoustic microscope (SAM) developed by Calvin Quate and coworkers in the 1970s, that made it possible to see structures at the nanoscale.

The tip of a scanning probe can also be used to manipulate nanostructures (a process called positional assembly). Feature-oriented scanning-positioning methodology suggested by Rostislav Lapshin appears to be a promising way to implement these nanomanipulations in automatic mode.

However, this is still a slow process because of low scanning velocity of the microscope. Various techniques of nanolithography such as dip pen nanolithography, electron beam lithography or nanoimprint lithography were also developed.

Lithography is a top-down fabrication technique where a bulk material is reduced in size to nanoscale pattern. The top-down approach anticipates nanodevices that must be built piece by piece in stages, much as manufactured items are made. Scanning probe microscopy is an important technique both for characterization and synthesis of nanomaterials. Atomic force microscopes and scanning tunneling microscopes can be used to look at surfaces and to move atoms around.

By designing different tips for these microscopes, they can be used for carving out structures on surfaces and to help guide self-assembling structures. By using, for example, feature-oriented scanning-positioning approach, atoms can be moved around on a surface with scanning probe microscopy techniques. At present, it is expensive and time-consuming for mass production but very suitable for laboratory experimentation.

In contrast, bottom-up techniques build or grow larger structures atom by atom or molecule by molecule. These techniques include chemical synthesis, self-assembly and positional assembly. Another variation of the bottom-up approach is molecular beam epitaxy or MBE. Researchers at Bell Telephone Laboratories like John R. Arthur. Alfred Y. Cho, and Art C. Gossard developed and implemented MBE as a rcscarch tool in the late 1960s and 1970s. Samples made by MBE were key to the discovery of the fractional quantum Hall effect for which the 1998 Nobel Prize in Physics was awarded. MBE allows scientists to lay down atomically-precise layers of atoms and, in the process, build up complex structures. Important for research on semiconductors, MBE is also widely used to make samples and devices for the newly emerging field of spintronics.[25]

Newer techniques such as Dual Polarisation Interferometry are enabling scientists to measure quantitatively the molecular interactions that take place at the nano-scale. However, new therapeutic products, based on responsive nanomaterials, such as the ultradeformable, stress-sensitive Transfersome vesicles, are under development and already approved for human use in some countries.[*citation needed*]

APPLICATIONS OF NANOTECHNOLOGY

With nanotechnology, a large set of materials and improved products rely on a change in the physical properties when the feature sizes are shrunk. Nanoparticles for example take advantage of their dramatically increased surface area to volume ratio. Their optical properties, *e.g.* fluorescence, become a function of the particle diameter. When brought into a bulk material, nanoparticles can strongly influence the mechanical properties of the material, like stiffness or elasticity. For example, traditional polymers can be reinforced by nanoparticles resulting in novel materials which can be used as lightweight replacements for metals.

Therefore, an increasing societal benefit of such nanoparticles can be expected. Such nanotechnologically enhanced materials will enable a weight reduction accompanied by an increase in stability and an improved functionality. There are many applications of nano technology, few of them are shown here.

MEDICINE

The biological and medical research communities have exploited the unique properties of nanomaterials for various applications (*e.g.*, contrast agents for cell imaging and therapeutics for treating cancer). Terms such as *biomedical nanotechnology*, *bionanotechnology*, and *nanomedicine* arc uscd to dcscribc this hybrid field. Functionalities can be added to nanomaterials by interfacing them with biological molecules or structures. The size of nanomaterials is similar to that of most biological molecules and structures; therefore, nanomaterials can be useful for both in vivo and in vitro biomedical research and applications. Thus far, the integration of nanomaterials with biology has led to the development of diagnostic devices, contrast agents, analytical tools, physical therapy applications, and drug delivery vehicles.

Diagnostics

Nanotechnology-on-a-chip is one more dimension of lab-on-a-chip technology. Biological tests measuring the presence or activity of selected substances become quicker, more sensitive and more flexible when certain nanoscale particles are put to work as tags or labels. Magnetic nanoparticles, bound to a suitable antibody, are used to label specific molecules, structures or microorganisms. Gold nanoparticles tagged with short segments of DNA can be used for detection of genetic sequence in a sample.

Multicolour optical coding for biological assays has been achieved by embedding different-sized quantum dots into polymeric microbeads. Nanopore technology for analysis of nucleic acids converts strings of nucleotides directly into electronic signatures.

Drug Delivery

The overall drug consumption and side-effects can be lowered significantly by depositing the active agent in the morbid region only and in no higher dose than needed. This highly selective approach reduces costs and human suffering. An example can be found in dendrimers and nanoporous materials.

They could hold small drug molecules transporting them to the desired location. Another vision is based on small electromechanical systems; NEMS are being investigated for the active release of drugs. Some potentially important applications include cancer treatment with iron nanoparticles or gold shells.

A targeted or personalized medicine reduces the drug consumption and treatment expenses resulting in an overall societal benefit by reducing the costs to the public health system. Nanotechnology is also opening up new opportunities in implantable delivery systems, which are often preferable to the use of injectable drugs, because the latter frequently display first-order kinetics (the blood concentration goes up rapidly, but drops exponentially over time). This rapid rise may cause difficulties with toxicity, and drug efficacy can diminish as the drug concentration falls below the targeted range.

Tissue Engineering

Nanotechnology can help to reproduce or to repair damaged tissue. This so called "tissue engineering" makes use of artificially stimulated cell proliferation by using suitable nanomaterial-based scaffolds and growth factors. Tissue engineering might replace today's conventional treatments like organ transplants or artificial implants. On the other hand, tissue engineering is closely related to the ethical debate on human stem cells and its ethical implications.

CHEMISTRY AND ENVIRONMENT

Chemical catalysis and filtration techniques are two prominent examples where nanotechnology already plays a role. The synthesis provides novel materials with tailored features and chemical properties: for example, nanoparticles with a distinct chemical surrounding (ligands), or specific optical properties. In this sense, chemistry is indeed a basic nanoscience. In a short-term perspective, chemistry will provide novel "nanomaterials" and in the long run, superior processes such as "self-assembly" will enable energy and time preserving strategies. In a sense, all chemical synthesis can be understood in terms of nanotechnology, because of its ability to manufacture certain molecules. Thus, chemistry forms a base for nanotechnology providing tailor-made molecules, polymers, etcetera, as well as clusters and nanoparticles.

CATALYSIS

Chemical catalysis benefits especially from nanoparticles, due to the extremely large surface to volume ratio. The application potential of nanoparticles in catalysis ranges from fuel cell to catalytic converters and photocatalytic devices. Catalysis is also important for the production of chemicals. Platinum nanoparticles are now being considered in the next generation of automotive catalytic converters because the very high surface area of nanoparticles could reduce the amount of platinum required. However, some concerns have been raised due to experiments demonstrating that they will spontaneously combust if methane is mixed with the ambient air. Ongoing research at the Centre National de la Recherche Scientifique (CNRS) in France may resolve their true usefulness for catalytic applications. Nanofiltration may come to be an important application, although future research must be careful to investigate possible toxicity.

Reduction of Energy Consumption

A reduction of energy consumption can be reached by better insulation systems, by the use of more efficient lighting or combustion systems, and by use of lighter and stronger materials in the transportation sector.

Currently used light bulbs only convert approximately 5 per cent of the electrical energy into light. Nanotechnological approaches like light-emitting diodes (LEDs) or quantum caged atoms (QCAs) could lead to a strong reduction of energy consumption for illumination.

Increasing the Efficiency of Energy Production

Today's best solar cells have layers of several different semiconductors stacked together to absorb light at different energies but they still only manage to use 40 per cent of the Sun's energy. Commercially available solar cells have much lower efficiencies (15-20 per cent).

Nanotechnology could help increase the efficiency of light conversion by using nanostructures with a continuum of bandgaps. The degree of efficiency of the internal combustion engine is about 30-40 per cent at the moment.

Nanotechnology could improve combustion by designing specific catalysts with maximized surface area. In 2005, scientists at the University of Toronto developed a spray-on nanoparticle substance that, when applied to a surface, instantly transforms it into a solar collector.

The Use of More Environmentally Friendly Energy Systems

An example for an environmentally friendly form of energy is the use of fuel cells powered by hydrogen, which is ideally produced by renewable energies. Probably the most prominent nanostructured material in fuel cells is the catalyst consisting of carbon supported noble metal particles with diameters of 1-5 nm.

Suitable materials for hydrogen storage contain a large number of small nanosized pores. Therefore many nanostructured materials like nanotubes, zeolites or alanates are under investigation. Nanotechnology can contribute to the further reduction of combustion engine pollutants by nanoporous filters, which can clean the exhaust mechanically, by catalytic converters based on nanoscale noble metal particles or by catalytic coatings on cylinder walls and catalytic nanoparticles as additive for fuels.

Recycling of Batteries

Because of the relatively low energy density of batteries the operating time is limited and a replacement or recharging is needed. The huge number of spent batteries and accumulators represent a disposal problem.

The use of batteries with higher energy content or the use of rechargeable batteries or supercapacitors with higher rate of recharging using nanomaterials could be helpful for the battery disposal problem.

THE EMERGING SCIENCE OF ENVIRONMENTAL APPLICATIONS

The science of earth and environment has matured through two major phases and is entering a third. In the first phase, which ended two decades ago, Earth and environmental science was largely discipline oriented and focused on developing knowledge in geology, atmospheric chemistry, ecosystems, and other aspects of the Earth system. In the 1980s, the scientific community recognized the close coupling of these disciplines and began to study them as interacting elements of a single system.

During this second phase, the paradigm of Earth system science emerged. With it came the ability to understand complex, system-oriented phenomena such as climate change, which links concepts from atmospheric sciences, biology, and human behaviour. Essential to the study of Earth's interacting systems was the ability to acquire, manage, and make available data from satellite observations; in parallel, new models were developed to express our growing understanding of the complex processes in the dynamic Earth system.

In the emerging third phase, knowledge developed primarily for the purpose of scientific understanding is being complemented by knowledge created to target practical decisions and action. This new knowledge endeavor can be referred to as the science of environmental applications. Climate change provides the most prominent example of the importance of this shift. Until now, the climate science community has focused on critical questions involving basic knowledge, from measuring the amount of change to determining the causes.

With the basic understanding now well established, the demand for climate applications knowledge is emerging. How do we quantify and monitor total forest biomass so that carbon markets can characterize supply? What are the implications of regional shifts in water resources for demographic trends, agricultural output, and energy production?

To what extent will seawalls and other adaptations to rising sea level impact coasts? These questions are informed by basic science, but they raise additional issues that can be addressed only by a new science discipline focused specifically on applications- a discipline that integrates physical, biogeochemical, engineering, and human processes. Its principal questions reflect a fundamental curiosity about the nature of the world we live in, tempered by the awareness that a question's importance scales with its relevance to a societal imperative. As Nobel laureate and U.S.

Secretary of Energy Steven Chu has remarked, "We seek solutions. We don't seek- dare I say this?-just scientific papers anymore". To illustrate the relationships between basic science and applications, consider the role of snowmelt runoff in water supplies. Worldwide, 1 billion people depend on snow or glacier melt for their water resources. Design and operations of water systems have traditionally relied on historical measurements in a stationary climate, along with empirical relationships and models. As climates and land use change, populations grow and relocate, and our built systems age and decay, these empirical methods of managing our water become inaccurate-a conundrum characterized as "stationarity is dead".

Snowmelt commonly provides water for competing uses: urban and agricultural supply, hydropower, recreation, and ecosystems. In many areas, both rainfall and snowfall occur, raising the concern that a future warmer climate will lead to a greater fraction of precipitation as rain, with the water arriving months before agricultural demand peaks and with more rapid runoff leading to more floods. In these mixed rain and snow systems, the societal need is: How do we sustain flood control and the benefits that water provides to humans and ecosystems when changes in the timing and magnitude of runoff are likely to render existing infrastructure inadequate?

The solution to the societal need requires a more fundamental, process-based understanding of the water cycle. Currently, historical data drive practices and decisions for flood control and water supply systems. Flood operations and reservoir flood capacity are predetermined by regulatory orders that are static, regardless of the type of water year, current state of the snowpack, or risk of flood. In many years, early snowmelt is not stored because statistically based projections anticipate floods that better information might suggest cannot materialize because of the absence of snow. The more we experience warming, the more frequently this occurrence will impact the water supply.

The related science challenges are:

- The statistical methods in use do not try to estimate the basin's water balance, and with the current measurement networks even in the U.S., we lack adequate knowledge of the amount of snow in the basins;
- We are unable to partition the input between rain and snow, or to partition that rain or snow between evapotranspiration and runoff;
- We lack the knowledge to manage the relationship between snow cover, forests, and carbon stocks;
- Runoff forecasts that are not based on physical principles relating to snowmelt are often inaccurate;
- We do not know what incentives and institutional arrangements would lead to better management of the watershed for ecosystem services.

Generally, models do not consider these kinds of interactions; hence the need for a *science of environmental applications.*

Its core characteristics differentiate it from the basic science of Earth and environment:

- *Need driven versus curiosity driven*: Basic science is question driven; in contrast, the new applications science is guided more by societal needs than scientific curiosity. Rather than seeking answers to questions, it focuses on creating the ability to seek courses of action and determine their consequences.
- *Externally constrained*: External circumstances often dictate when and how applications knowledge is needed. The creation of carbon trading markets will not wait until we fully quantify forest carbon content. It will happen on a schedule dictated by policy and economics. Construction and repair of the urban water infrastructure will not wait for an understanding of evolving rainfall patterns. Applications science must be prepared to inform actions subject to these external drivers, not according to academic schedules based on when and how the best knowledge can be obtained.
- *Consequential and recursive*: Actions arising from our knowledge of the Earth often change the Earth, creating the need for new knowledge about what we have changed. For example, the more we knew in the past about locations of fish populations, the more the populations were overfished; our original knowledge about them became rapidly outdated through our own actions. Applications science seeks to understand not just those aspects of the Earth addressed by a particular use scenario, but also the consequences and externalities that result from that use scenario. A recent example is the shift of agricultural land to corn-for ethanol production—an effort to reduce

climate change that we now recognize as significantly stressing scarce water resources.

- *Useful even when incomplete*: As the snowpack example illustrates, actions are often needed despite incomplete data or partial knowledge. The difficulty of establishing confidence in the quality of our knowledge is particularly disconcerting given the loss of stationarity associated with climate change. New means of making effective use of partial knowledge must be developed, including robust inference engines and statistical interpretation.
- *Scalable*: Basic science knowledge does not always scale to support applications needs. The example of carbon trading presents an excellent illustration.
- Basic science tells us how to relate carbon content to measurements of vegetation type and density, but it does not give us the tools that scale this to a global inventory. New knowledge tools must be built to accurately create and update this inventory through cost-effective remote sensing or other means.
- *Robust*: The decision makers who apply applications knowledge typically have limited comprehension of how the knowledge was developed and in what situations it is applicable. To avoid misuse, the knowledge must be characterized in highly robust terms. It must be stable over time and insensitive to individual interpretations, changing context, and special conditions.
- *Data intensive*: Basic science is data intensive in its own right, but data sources that support basic science are often insufficient to support applications. Localized impacts with global extent, such as intrusion of invasive species, are often difficult for centralized projects with small numbers of researchers to ascertain.

New applications-appropriate sources must be identified, and new ways of observing (including the use of communities as data gatherers) must be developed. Each of these characteristics implies development of *new knowledge types* and *new tools for acquiring that knowledge*. The snowpack example illustrates what this requirement means for a specific application area.

Four elements have recently come together that make deployment of a measurement and information system that can support decisions at a scale of a large river basin feasible:

1. Accurate, sustained satellite estimates of snow-covered area across an entire mountain range;
2. Reliable, low-cost sensors and telemetry systems for snow and soil moisture;
3. Social science data that complement natural and engineered systems data to enable analysis of human decision making;

4. Cyberinfrastructure advances to integrate data and deliver them in near real time.

For snow-dominated drainage basins, the highest-priority scientific challenge is to estimate the spatial distribution and heterogeneity of the *snow water equivalent*— *i.e.*, the amount of water that would result if the snow were to melt. Because of wind redistribution of snow after it falls, snow on the ground is far more heterogeneous than rainfall, with several meters of differences within a 10 to 100 m distance.

Heterogeneity in snow depth smoothes the daily runoff because of the variability of the duration of meltwater in the snowpack; seasonally, it produces quasi-riparian zones of increased soil moisture well into the summer.

The approach to estimating the snow water equivalent involves several tasks using improved data:

- Extensive validation of the satellite estimates of snow cover and its reflectivity;
- Using results from an energy balance reconstruction of snow cover to improve interpolation from more extensive ground measurements and satellite data;
- Development of innovative ways to characterize heterogeneity;
- Testing the interpolated estimates with a spatially distributed runoff model. The measurements would also help clarify the accuracy in precipitation estimates from regional climate models. This third phase of Earth and environmental science will evolve over the next decade as the scientific community begins to pursue it. Weather science has already built substantial capability in applications science; the larger field of Earth science will need to learn from and extend those efforts. The need for basic science and further discovery will not diminish, but instead will be augmented and extended by this new phase. The questions to address are both practically important and intellectually captivating. Will our hydrologic forecasting skill decline as changes in precipitation diminish the value of statistics obtained from historic patterns? Where will the next big climate change issue arise, and what policy actions taken today could allow us to anticipate it? Equally important is improving how we apply this knowledge in our daily lives. The Internet and mobile telephones, with their global reach, provide new ways to disseminate information rapidly and widely. Information was available to avoid much of the devastation from the Asian tsunami and Hurricane Katrina, but we lacked the tools for rapid decision making and communication of needed actions. Applications science is therefore integrative; it couples understanding of physical phenomena and research into the ways that people and organizations can use better knowledge to make decisions. The public

as a whole can also become an important contributor to localized Earth observation, augmenting our limited satellite and sensor networks through devices as simple as mobile phone cameras. The ability to leverage this emerging data-gathering capability will be an important challenge for the new phase of environmental science. The security and prosperity of nearly 7 billion people depend increasingly on our ability to gather and apply information about the world around us. Basic environmental science has established an excellent starting point. We must now develop this into a robust science of environmental applications.

GLOBAL ENVIRONMENTAL PROBLEMS

Ethics is understood to be the domain of enquiry that explores what is right or wrong, obligatory or non-obligatory, or when responsibility attaches to human Behaviour. Are there features of global environmental problems that call for classifying them as essentially ethical problems with even greater force than some local or regional environmental problems? If so, what are these features?

Why is this important? If some global environmental problems are essentially ethical problems, then some of the excuses that nations often use to justify not reducing their contributions to the global environmental problems are ethically problematic. This is so because ethical obligations entail duties and responsibilities to others with the result that national policies may not be justified on national interests alone. That is, if there are ethical obligations to others to cease causing harm, then policy options must be responsive to these obligations. The features of global environmental problems that strongly call for classification as ethical problems. Some global environmental problems are caused by people in one part of the world but are most harshly experienced by others who are separated from those causing the problem by great space and time.

In addition, those most vulnerable to global environ-mental problems are often least responsible for causing them. Climate change is a strong example of this feature of global environmental problems. Those most vulnerable to climate include many of the poorest people in the world who are among those least responsible for emitting the greenhouse gases that are causing the problem. The same separation between victim and cause is also true of other global environmental problems, including upper atmospheric ozone depletion, some ocean degradation problems including destruction of global fisheries, some biodiversity and forest problems, and exposure to toxic substances that are deposited around the world by long-range air transport.

In the case of global fisheries depletion, for instance, fishing trawlers from some countries often go great distances to harvest fish for domestic consumption, yet the harshest consequences of depleted global fisheries are often experienced by subsistence fisherman in poor countries far away from the home port of the fishing boats.

This separation of space and time and between cause and impact of course can also happen in more local environmental problems such as water pollution, but the distances between those causing the problem and those who suffer the consequences are most problematic at the global scale because of the magnitude of the distances at this scale.. With the rise in truly global environmental problems in the last few decades, it is now abundantly clear that Behaviour of people in one part of the world can gravely affect the human health, economic viability and overall welfare of others far away, but their suffering is almost invisible to those who cause the problem.

Several implications for ethical reasoning follow directly from this separation of distance and time between cause and effect. First, our intuitive sense of ethical responsibility for harms that we cause others is harder to trigger when those who will be harmed by our action are people we will not ever know and are from a different culture. Yet ethics and morality require us to refrain from harming all others who may suffer because of our Behaviour even if they are separated from us in time and space.

Secondly, because the actual harm that we are causing to others may be in a particular part of the world that is uniquely vulnerable to global environmental change but is unknown to us, we not only don't know the people we are harming but have no idea where the locus of the worst harm is located. For instance, particularly vulnerable villages to climate change impacts may turn on the geographical features where the village is located such as dependence on glacier water flows in the summer coupled with vulnerability to monsoon irregularities. Yet few of us know where these places are. And so in the case of global environmental problems we may be causing great harm to others who we not only do not know personally, but about whom we neither have an idea of their location nor why they are particularly vulnerable to actions we are taking. For this reason, global scale environmental problems create a huge challenge to trigger our sense of ethical responsibility, yet ethics requires us to not harm others through our actions none the less.

A related reason why the global environmental problems demand classification as ethical problems stems from the fact that some actions that cause some environmental damage are often justified on the basis of cost-benefit analysis. Yet this justification is ethically problematic when those who benefit from an action that causes harm have no relation with those who are harmed. For instance, government may tolerate inevitable pollution from a land fill by identifying the benefits that citizens receive by having a place to take their garbage. But in the case of global environmental problems, those who are harmed by actions do not enjoy the benefits that come from taking the action.

A harms-benefits justification for environmental degradation, if ethically justifiable at all, only works where those who will suffer the harms and those who benefit from the action that causes the harm are in the same community of interests. In the case of global environmental problems, those who are harmed are rarely made better by the benefits of the action that cause the harm.

The second reason why global environmental problems call for classification as ethical problems stems from the fact that their consequences are often catastrophic to those who are most harmed by them. Climate change, for instance, directly threatens human life and health and resources to sustain life as well as species of plants and animals and ecosystems around the world. The harms include deaths from disease, droughts, floods, heat and intense storms and damage to homes and villages from rising oceans and intense storms, adverse impacts on agriculture, social disputes caused by diminishing natural resources, sickness from a variety of diseases, the inability to rely upon traditional sources of food, the destruction of water supplies, and the inability to live where one has lived to sustain life. In addition, the very existence of some small island nations is threatened by climate change caused seal level rise. Clearly these impacts are catastrophic for some.

Loss of global fisheries can be devastating to substance fishermen, global desertification and drought are likely to cause mass starvation, loss of atmospheric ozone can cause skin cancer in millions, and long-range transport of persistent organic chemicals can cause deadly diseases and birth defects. Because ethics requires people to be particularly careful to not to harm others when the harm is great, global environmental problems need to be seen as ethical issues.

Governments are expected to protect their citizens from life-threatening dangers. At the local, regional or national scale, citizens can petition their governments to protect them from environmental harms. But at the global level, no government exists whose jurisdiction matches the scale of global environmental problems. And so, although existing national, regional and local governments have jurisdiction over activities within their boundaries coupled to responsibilities to their citizens, they have no responsibility to those outside their boarders in the absence of binding international law. For this reason, ethical appeals to get governments to require that their citizens not harm non-citizens in other nations are necessary to prompt action.

In fact, according to a leading foreign policy theory called "realism," national governments should limit the scope of their international engagements to the pursuit of national interest alone. According to some "realist" proponents, governments should refrain from acting at the international level for ethical reasons that do not coincide with national interests. Yet global environmental problems require people in one part of the world to consider the interests of others outside the jurisdiction of their national governments. Because there is no duty of national governments to protect foreigners from global environmental problems unless they bind themselves voluntarily in treaties, an appeal to ethical and moral responsibility is particularly important to motivate national actions in regard to global scale environmental problems. Moreover, if global environmental problems actually create ethical obligations, they have ethical duties to mitigate global environmental problems

CFCS IN ATMOSPHERE

There has been a movement since the late 1970s to ban CFCs because of their destructive effect on the ozone layer. This damage was discovered by Sherry Rowland and Mario Molina, who first published a paper suggesting the connection in 1974. It turns out that one of CFCs' most attractive features-their unreactivity-has been instrumental in making them one of the most significant pollutants. CFCs' lack of reactivity gives them a lifespan which can exceed 100 years in some cases. This gives them time to diffuse into the upper stratosphere. Here, the sun's ultraviolet radiation is strong enough to break off the chlorine atom, which on its own is a highly reactive free radical. This catalyses the break up of ozone into oxygen:

$$Cl + O_3 \rightarrow ClO + O_2$$

$$ClO + O \rightarrow Cl + O_2$$

CFCs are a problem because the chlorine is regenerated at the end of these reactions, making it able to repeat this reaction millions of times. This disrupts the ionic structure of oxygen atoms, causing their lowest ionisation energy to drop. This reaction is believed to be the cause of the ozone hole observed over the poles and upper latitudes of the Earth. Decreases in stratospheric ozone have been linked to increases in skin cancer. In 1975, Oregon enacted the world's first ban of CFCs. By 1985, scientists observed a dramatic seasonal depletion of the ozone layer over Antarctica, which led the United Nations to take action in 1987. International attention to CFCs resulted in a meeting of world diplomats in Montreal in 1987. They forged a treaty, the Montreal Protocol, which called for drastic reductions in the production of CFCs. On March 2, 1989, 12 European Community nations agreed to ban the production of all CFCs by the end of the century. In 1990, diplomats met in London and voted to significantly strengthen the Montreal Protocol by calling for a complete elimination of CFCs by the year 2000. By the year 2010 CFCs should be completely eliminated from developing countries as well

FREON

Freon is a trade name for a group of chlorofluorocarbons used primarily as a refrigerant. The word Freon is a registered trademark belonging to Dupont. Prior to their implication as a major contributor to ozone depletion (and a minor contributor to the greenhouse effect), freons were also used as aerosol propellants and to foam up polymer materials.

Freon was developed as a safer alternative to toxic gases such as, ammonia (NH_3), methyl chloride (CH_3Cl), and sulfur dioxide (SO_2), that were used as refrigerants prior to the discovery of freon. Despite the controversy, R12 works well and is very safe. Its inventor, Thomas Midgley, actually breathed in a lungful and exhaled over a candle to demonstrate its non-toxicity and non-flammability. Before CFCs were banned by developed nations, a 12 ounce (340 g) can of freon (R12) cost about $1; following the ban the cost of freon rose dramatically. People now go through considerable trouble to recycle freon from existing sources. As of 2002, freon cost about $85 per lb ($187 per kg). Freon is most frequently sold in 12 ounce (340 g) cans and 30 pound (13.6 kg) canisters.

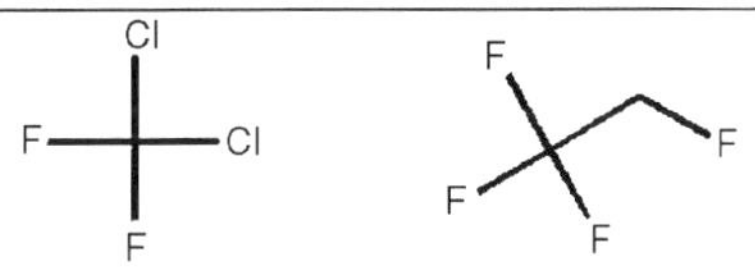

Dichlordifuoromethane "Freon 12" 1,1,1,2-Tetrafluoroethane "Suva 134a"

Freon 12 → (also known as R–12) was a popular re-frigerant before it was banned. The chemical formula is $CCl_2F_2$and its proper chemical name is dichloro-difluoromethane. When people refer to freon without specifying a number, it is likely that they mean freon-12.

Suva 134a → (also known as R–134 or R–134a) is the most common replacement for R–12. Unlike R–12, which is a chlorofluorocarbon, R–134 is a hydro fluorocarbon. The chemical formula is CH_2FCF_3 and the proper chemical name 1,1,1,2-tetrafluoroethane.

	Freon 41	**Fluoromethane**
R-11	CFC-11	Trichlorofluoromethane
R-12	CFC-12	Dichlorodifluoromethane
R-13	CFC-13	Chlorotrifluoromethane
R-22	HCFC-22	Chlorodifluoromethane
R-23	HFC-23	Trifluoromethane
R-113	CFC-113	Trichlorotrifluoroethane
R-114	CFC-114	1,2-dichloro-1,1,2,2-Tetrafluoroethane
R-115	CFC-115	1-chloro-1,1,2,2,2-Pentafluoroethane
R-116	CFC-116	Hexachloroethane
R-134a	HFC-134a	1,1,1,2-tetrafluoroethane
R-227ea	HFC-227ea	1,1,1,2,3,3,3-heptafluoropropane

Freon in copper tubing open to the environment can turn into phosgene gas after coming in contact with extreme heat, such as, while brazing or in a fire situation. Phosgene is a substance that was used as a chemical weapon in

World War I. Low exposure can cause irritation, but high levels cause fluid to collect in the lungs, possibly resulting in death.

HALON

Halons are a group of compounds obtained by replacing the hydrogen atoms of a hydrocarbon with halogen atoms, such as bromine or fluorine. Halon 1211 is bromochlo-rodifluoromethane (CF_2BrCl) and Halon 1301 is bromotrifl-uoromethane (CF_3Br). Halons are very stable and are widely used in fire extinguishers where water and other alternatives would be ineffective and dangerous (*e.g.* when dealing with oil fires) or cause unacceptable collateral damage (*e.g.* with electronic equipment.)

Introduction

At high temperatures, halons decompose to release halogen atoms that combine readily with active hydrogen atoms, quenching the flame propagation reaction even when adequate fuel, oxygen and heat remains. The chemical reaction in a flame proceeds as a free radical chain reaction; by sequestering the radicals which propagate the reaction, halons are able to "poison" the fire at much lower concentrations than are required by fire suppressants using the more traditional methods of cooling, oxygen deprivation, or fuel dilution. For example, Halon 1301 flood systems are typically used at concentrations no higher than 6 per cent v/v in air, and can suppress many fires at 2.9 per cent v/v. By contrast, carbon dioxide fire suppression flood systems are operated from 34 per cent (surface-only combustion of liquid fuels) up to 75 per cent (dust traps). Carbon dioxide can cause severe distress at concentrations of 3 to 6 per cent, and has caused death by respiratory paralysis in a few minutes at 10 per cent concentration.

Halon 1301 causes only slight giddiness at its effective concentration of 5 per cent, and even at 15 per cent persons remain conscious but impaired, and suffer no long term effects. (Experimental animals have also been exposed to 2 per cent concentrations of Halon 1301 for 30 hours per week for 4 months, with no discernible health effects at all.) Halon 1211 also has low toxicity, although it is more toxic than Halon 1301, and thus considered unsuitable for flooding systems.

Fig. Halon Canisters Used in a Fire-suppression System

However, halon fire suppression is not completely non-toxic; very high temperature flame, or contact with red-hot metal, can cause decomposition to toxic byproducts. The presence of such byproducts is readily detected because they include hydrobromic acid and hydrofluoric acid, which are intensely irritating. Halons are very effective on Class A (organic solids), B (flammable liquids and gases) and C (electrical) fires, but they are totally unsuitable for Class D (metal) fires, as they will not only produce toxic gas and fail to halt the fire, but in some cases pose a risk of explosion. Halons can be used on Class K (kitchen oils and greases) fires, but offer no advantages over specialised foams.

Nomenclature

Halons are referred to by a system of code numbers similar to (but simpler than) the system used for freons. The first digit specifies the number of carbon atoms in the molecule, the second is the number of fluorine atoms, the third is the chlorine atoms, and the fourth is the number of bromine atoms. If the number includes a fifth digit, the fifth number indicates the number of iodine atoms (though iodine in Halon is rare). (Note that the various atomic species are listed in order of increasing atomic number.) Any bonds not taken up by halogen atoms are then allocated to hydrogen atoms.

For example, consider Halon 1211: C F Cl Br 1 2 1 1

Halon 1211 has one carbon atom, two fluorine atoms, one chlorine atom and one bromine atom. A single carbon only has four bonds, all of which are taken by the halogen atoms, so there is no hydrogen. Thus its formula is CF_2BrCl, and its IUPAC name is therefore bromochlorodi fluoromethane.

Application

Halon 1211 is typically used in hand-held extinguishers, in which a stream of liquid halon is directed at a smaller fire by a user. The stream evaporates under reduced pressure, producing strong local cooling, as well as a high concentration of halon in the immediate vicinity of the fire. In this mode, extinguishment is achieved by cooling and oxygen deprivation at the core of the fire, as well as radical quenching over a larger area. After fire suppression, the halon moves away with the surrounding air, leaving no residue.

Halon 1301 is more usually employed in flooding systems. In these systems, banks of halon cylinders are kept pressurised to about 4 MPa with compressed nitrogen, and ducts lead to the protected enclosure. On triggering, the entire measured contents of one or more cylinders are dumped into the enclosure in a few seconds, through nozzles designed to ensure uniform mixing throughout the room. The quantity dumped is pre-calculated to achieve the desired concentration, typically 3-5 per cent v/v. This level is maintained for some time, typically twenty minutes, to ensure all items have cooled so reignition is unlikely to occur, then the air in the enclosure is purged. During this time the enclosure may be entered by persons wearing SCBA. (There exists a common myth that this is because halon is highly toxic; in fact it is because it can cause giddiness and mildly impaired perception, and also due to the risk of combustion byproducts.) Flooding systems may be manually operated or automatically triggered by a VESDA. In the latter case, a warning siren and strobe lamp will first be activated for a few seconds to warn personnel to evacuate the area. The rapid dumping of halon and consequent rapid cooling fills the air with fog, and is accompanied by a loud, disorienting noise.

Environmental Issues

There is concern that halons are being broken down in the atmosphere to bromine, which reacts with ozone, leading to depletion of the ozone layer (this is similar to the case of chlorofluorocarbons such as freon). These issues are complicated: the kinds of fires that require halon extinguishers to be put out will typically cause more damage to the ozone layer than the halon itself, not to mention human and property damage. However, fire extinguisher systems must be tested regularly, and these tests may lead to damage. As a result, some regulatory measures have been taken, and halons are being phased out in most of the world.

History

Carbon tetrachloride was used in fire extinguishers and glass (anti)-"fire grenades"from the latenineteenth century until around the end of World War II. Experimentation with halogenated hydrocarbons for fire suppression on military aircraft began at least as early as the 1920s. During World War II, various early halons were in standard use in aircraft by some combatants, but these early halons suffered from excessive toxicity. Nevertheless after the war they slowly became more common in civil aviation as well.

In the 1960s, fluorinated halons became available, and were quickly recognised as one of the most effective fire fighting materials discovered. Much early research with Halon 1301 was conducted under the auspices of the US Armed Forces, while Halon 1211 was initially mainly developed in the UK. By the late 1960s, they were standard in many applications where water and dry powder extinguishers posed a threat of damage to the protected property, including computer rooms, telecommunications switches, Labouratories, museums and art collections. Beginning with warships in the 1970s, halons also progressively came to be associated with rapid knockdown of severe fires in confined spaces with minimal risk to personnel.

By the early 1980s, halons were in common use on aircraft, ships and large vehicles, as well as in computer facilities and galleries. However, concern was beginning to be felt about the possible impact of chlorinated and brominated hydrocarbons on the ozone layer. The Vienna Convention on Ozone Layer Protection did not cover halons, as it was felt that emergency discharge of systems was too small in volume to produce a significant impact, and too important to human safety for restriction. However, by the time of the Montreal

Protocol it was realised that discharges during system tests and maintenance accounted for substantially larger volumes than emergency discharges, and so halons were brought into the treaty, but with many exceptions. Halons have been generally phased out and the possession of halon equipment is prohibited in some countries like the Netherlands and Belgium from January 1, 2004, based on the Montreal Protocol and guidelines of the European Union. Production of new stocks has ceased in most (probably all) countries as of 1994. However many countries still require aircraft to be fitted with halon fire suppression systems, as no safe and completely satisfactory alternative has been discovered for this application. There are also a few other highly specialised users.

INFORMATION AND COMMUNICATION

Current high-technology production processes are based on traditional top down strategies, where nanotechnology has already been introduced silently. The critical length scale of integrated circuits is already at the nanoscale (50 nm and below) regarding the gate length of transistors in CPUs or DRAM devices.

Memory Storage

Electronic memory designs in the past have largely relied on the formation of transistors however research into crossbar switch based electronics have offered an alternative using reconfigurable interconnections between vertical and horizontal wiring arrays to create ultra high density memories. Two leaders in this area are Nantero which has developed a carbon nanotube based crossbar memory called Nano-RAM and Hewlett-Packard which has proposed the use of memristor material as a future replacement of Flash memory.

Novel Semiconductor Devices

An example of such novel devices is based on spintronics.The dependence of the resistance of a material (due to the spin of the electrons) on an external field is called magnetoresistance. This effect can be significantly amplified (GMR - Giant Magneto-Resistance) for nanosized objects, for example when two ferromagnetic layers are separated by a non-magnetic layer, which is several nanometers thick (*e.g.* Co-Cu-Co). The GMR effect has led to a strong increase in the data storage density of hard disks and made the gigabyte range possible. The so called tunneling magnetoresistance (TMR) is very similar to GMR and based on the spin dependent tunneling of electrons through adjacent ferromagnetic layers. Both GMR and TMR effects can be used to create a non-volatile main memory for computers, such as the so called magnetic random access memory or MRAM. In 1999, the ultimate CMOS transistor developed at the Laboratory for Electronics and Information Technology in Grenoble, France, tested the limits of the principles of the MOSFET transistor with a diameter of 18 nm (approximately 70 atoms placed side by side). This was almost one tenth the size of the smallest industrial transistor in 2003. It enabled the theoretical integration of seven billion junctions on a □1 coin. However, the CMOS transistor, which was created in 1999, was not a simple research experiment to study how CMOS technology functions, but rather a demonstration of how this technology functions now that we ourselves are getting ever closer to working on a molecular scale. Today it would be impossible to master the coordinated assembly of a large number of these transistors on a circuit and it would also be impossible to create this on an industrial level.

Novel Optoelectronic Devices

In the modern communication technology traditional analog electrical devices are increasingly replaced by optical or optoelectronic devices due to their enormous bandwidth and capacity, respectively. Two promising examples are photonic crystals and quantum dots. Photonic crystals are materials with a periodic variation in the refractive index with a lattice constant that is half the wavelength of the light used. They offer a selectable band gap for the propagation of a certain wavelength, thus they resemble a semiconductor, but for light or photons instead of electrons. Quantum dots are nanoscaled objects, which can be used, among many other things, for the construction of lasers. The advantage of a quantum dot laser over the traditional semiconductor laser is that their emitted wavelength depends on the diameter of the dot. Quantum dot lasers are cheaper and offer a higher beam quality than conventional laser diodes.

Displays

The production of displays with low energy consumption could be accomplished using carbon nanotubes (CNT). Carbon nanotubes be electrically conductive and due to their small diameter of several nanometers, they can be used as field emitters with extremely high efficiency for field emission displays (FED). The principle of operation resembles that of the cathode ray tube, but on a much smaller length scale.

Quantum Computers

Entirely new approaches for computing exploit the laws of quantum mechanics for novel quantum computers, which enable the use of fast quantum algorithms. The Quantum computer will have quantum bit memory space termed qubit for several computations at the same time.

HEAVY INDUSTRY

AEROSPACE

Lighter and stronger materials will be of immense use to aircraft manufacturers, leading to increased performance. Spacecraft will also benefit, where weight is a major factor. Nanotechnology thereby helps to reduce the size of equipments used. Hang gliders halve their weight while increasing their strength and toughness through the use of nanotech materials. Nanotech is lowering the mass of supercapacitors that will increasingly be used to give power to assistive electrical motors for launching hang gliders off flatland to thermal-chasing altitudes.

Refineries

Using nanotech applications, refineries producing materials such as steel and aluminium will be able to remove any impurities in the materials they create.

Vehicle Manufacturers

Much like aerospace, lighter and stronger materials will be useful for creating vehicles that are both faster and safer. Combustion engines will also benefit from parts that are more hard-wearing and more heat-resistant.

CONSUMER GOODS

Nanotechnology is already impacting the field of consumer goods, providing products with novel functions ranging from easy-to-clean to scratch-resistant. Modern textiles are wrinkle-resistant and stain-repellent; in the mid-term clothes will become "smart", through embedded "wearable electronics". Already in use are different nanoparticle improved products. Especially in the field of cosmetics, such novel products have a promising potential.

Foods

Nanotechnology can be applied in the production, processing, safety and packaging of food. A nanocomposite coating process could improve food packaging by placing anti-microbial agents directly on the surface of the coated film. Nanocomposites could increase or decrease gas permeability of different fillers as is needed for different products. They can also improve the mechanical and heat-resistance properties and lower the oxygen transmission rate. Research is being performed to apply nanotechnology to the detection of chemical and biological substances for sensing biochemical changes in foods.

Household

The most prominent application of nanotechnology in the household is self-cleaning or "easy-to-clean" surfaces on ceramics or glasses. Nanoceramic particles have improved the smoothness and heat resistance of common household equipment such as the flat iron.

Optics

The first sunglasses using protective and antireflective ultrathin polymer coatings are on the market. For optics, nanotechnology also offers scratch resistant surface coatings based on nanocomposites. Nano-optics could allow for an increase in precision of pupil repair and other types of laser eye surgery.

Textiles

The use of engineered nanofibres already makes clothes water- and stain-repellent or wrinkle-free. Textiles with a nanotechnological finish can be washed less frequently and at lower temperatures. Nanotechnology has been used to integrate tiny carbon particles membrane and guarantee full-surface protection from electrostatic charges for the wearer.

Cosmetics

One field of application is in sunscreens. The traditional chemical UV protection approach suffers from its poor long-term stability. A sunscreen based on mineral nanoparticles such as titanium dioxide offer several advantages. Titanium oxide nanoparticles have a comparable UV protection property as the bulk material, but lose the cosmetically undesirable whitening as the particle size is decreased.

APPLICATIONS OF NANOTUBE

The properties of carbon nanotubes have caused researchers and companies to consider using them in several fields. For example, because carbon nanotubes have the highest strength to weight ratio of any known material, researchers at NASA are combining nanotubes with other materials into composites that can be used to build lightweight spacecraft. Another property of nanotubes is that they can easily penetrate membrances such as cell walls.

In fact, nanotubes long, narrow shape make them look like miniature needles, so it makes sense that they can function like a needle at the cellular level. Medical researchers are using this property by attaching molecules that are attracted to cancer cells to nanotubes to deliver drugs directly to the diseased cells.

Another interesting property of nanotubes is that their electrical resistance changes significantly when other molecules attach themselves to the carbon atoms. Companies are using this property to develop sensors that can detect chemical vapors such as carbon monoxide or biological molecules.

These are just a few of the potential uses of carbon nanotubes The following survey of carbon nanotube applications introduces these and many other uses.

A Survey of Carbon Nanotube Applications Under Development

Researchers and companies are working to use carbon nanotubes in various fields. The list below introduces many of these uses. Strong, lightweight composites of carbon nanotubes and other materials that can be used to build lightweight spacecraft. Cables made from carbon nanotubes strong enough to be used for the Space Elevator to drastically reduce the cost of lifting people and materials into orbit. Taking advantage of nanotubes ability to enter cancer cells by attaching targeting molecules which have an affinity to cancer cells as well as anti-cancer drugs to the nanotubes which safety transports an anti-cancer drug through the bloodstream to the tumor.

Stronger bicycle components made by adding carbon nanotubes to a matrix of carbon fibres. Improve the healing process for broken bones by providing a carbon nanotube scaffold for new bone material to grow on. Sensors using carbon nanotube detection elements capable of detecting a range of chemical vapors. These sensors depend upon the fact that the resistance of a carbon nanotube changes in the presence of a chemical vapour.

IMPLICATIONS

Due to the far-ranging claims that have been made about potential applications of nanotechnology, a number of serious concerns have been raised about what effects these will have on our society if realized, and what action if any is appropriate to mitigate these risks.

There are possible dangers that arise with the development of nanotechnology. The Centre for Responsible Nanotechnology suggests that new developments could result, among other things, in untraceable weapons of mass destruction, networked cameras for use by the government, and weapons developments fast enough to destabilize arms races ("Nanotechnology Basics").

One area of concern is the effect that industrial-scale manufacturing and use of nanomaterials would have on human health and the environment, as suggested by nanotoxicology research.

Groups such as the Centre for Responsible Nanotechnology have advocated that nanotechnology should be specially regulated by governments for these reasons. Others counter that overregulation would stifle scientific research and the development of innovations which could greatly benefit mankind.

Other experts, including director of the Woodrow Wilson Center's Project on Emerging Nanotechnologies David Rejeski, have testified[28] that successful commercialization depends on adequate oversight, risk research strategy, and public engage-ment. More recently local municipalities have passed (Berkeley, CA)[29] or are considering (Cambridge, MA)[30] - ordinances requiring nanomaterial manufacturers to disclose the known risks of their products.

The National Institute for Occupational Safety and Health is conducting research on how nanoparticles interact with the body's systems and how workers might be exposed to nano-sized particles in the manufacturing or industrial use of nanomaterials. NIOSH offers interim guidelines for working with nanomaterials consistent with the best scientific knowledge.[31]

In "The Consumer Product Safety Commission and Nanotechnology,"[32] E. Marla Felcher suggests that the Consumer Product Safety Commission, which is charged with protecting the public against unreasonable risks of injury or death associated with consumer products, is ill-equipped to oversee the safety of complex, high-tech products made using nanotechnology.

Longer-term concerns centre on the implications that new technologies will have for society at large, and whether these could possibly lead to either a post scarcity economy, or alternatively exacerbate the wealth gap between developed and developing nations.

Health and Environmental Concerns

Some of the recently developed *nanoparticle* products may have unintended consequences. Researchers have discovered that silver nanoparticles used in socks to reduce foot odour are being released in the wash with possible negative consequences.[33] Silver nanoparticles, which are bacteriostatic, may then destroy beneficial bacteria which are important for breaking down organic matter in waste treatment plants or farms.[34]

A study at the University of Rochester found that when rats breathed in nanoparticles, the particles settled in the brain and lungs, which led to significant increases in biomarkers for inflammation and stress response.[35]

A major study published more recently in Nature nanotechnology suggests some forms of carbon nanotubes – a poster child for the "nanotechnology revolution" – could be as harmful as asbestos if inhaled in sufficient quantities. Anthony Seaton of the Institute of Occupational Medicine in Edinburgh, Scotland, who contributed to the article on carbon nanotubes said "We know that some of them probably have the potential to cause mesothelioma.

So those sorts of materials need to be handled very carefully." [36]. In the absence of specific nano-regulation forthcoming from governments, Paull and Lyons (2008) have called for an exclusion of engineered nanoparticles from organic food.[37]

Regulation

Calls for tighter regulation of nanotechnology have occurred alongside a growing debate related to the human health and safety risks associated with nanotechnology. Further, there is significant debate about who is responsible for the regulation of nanotechnology.

While some non-nanotechnology specific regulatory agencies currently cover some products and processes (to varying degrees) – by "bolting on" nanotechnology to existing regulations – there are clear gaps in these regimes.[38] In "Nanotechnology Oversight: An Agenda for the Next Administration,"[39] former EPA deputy administrator J. Clarence (Terry) Davies lays out a clear regulatory roadmap for the next presidential administration and describes the immediate and longer term steps necessary to deal with the current shortcomings of nanotechnology oversight. Stakeholders concerned by the lack of a regulatory framework to assess and control risks associated with the release of nanoparticles and nanotubes have drawn parallels with bovine spongiform encephalopathy ('mad cow's disease), thalidomide, genetically modified food,[40] nuclear energy, reproductive technologies, biotechnology, and asbestosis.

The Royal Society report[42] identified a risk of nanoparticles or nanotubes being released during disposal, destruction and recycling, and recommended that "manufacturers of products that fall under extended producer responsibility regimes such as end-of-life regulations publish procedures outlining how these materials will be managed to minimize possible human and environmental exposure" (p.xiii). Reflecting the challenges for ensuring responsible life cycle regulation, the Institute for Food and Agricultural Standards has proposed standards for nanotechnology research and development should be integrated across consumer, worker and environmental standards.

They also propose that NGOs and other citizen groups play a meaningful role in the development of these standards. The term "manufacturers" includes persons and businesses that produce nanotubes in California, or import carbon nanotubes into California for sale. The purpose of this information request will be to identify information gaps and to develop information about carbon nanotubes, an important emerging nanomaterial.

SEMICONDUCTOR FABRICATION

MOORE'S LAW, OPTICAL LITHOGRAPHY AND THE SEARCH FOR ALTERNATIVES.

Computer chips (and the silicon based transistors within them) are rapidly shrinking according to a predictable formula. According to the Semiconductor Industry Association's extrapolation of formulas such as this one (SIA roadmap), it is expected that the sizes of circuits within our chips will reach the size of only a few atoms in about 20 years.

Since almost all of our modern computers are made from silicon semiconductor transistors patterned and carved by light (photolithography), the shrinking of circuits predicted by the SIA may not be the most economical method for the future. An enormous amount of money has been invested in the semiconductor industry in order to consistently shrink and improve our semiconductor electronics.

Smaller circuits require less energy, operate more quickly and, of course, take up less space. Thus, Moore's law has been adhered to since computers first became commercially available. However, this simple shrinking of components can not continue for much longer.

As transistors such as the Metal-Oxide Semiconductor Field Effect Transistor is made smaller, both its properties and manufacturing expense change with the scale. Currently, Ultraviolet light is used to create the silicon circuits with a lateral resolution around 200 nm (the wavelength of ultraviolet light). As the circuits shrink below 100 nm new fabrication methods must be created, resulting in increasing costs.

Furthermore, once the circuit size reaches only a few nanometers, quantum effects such as tunneling begin to become important, which drastically changes the ability for the computers to function normally. Thus, novel methods for computer chip fabrication have been and are being intensely sought by microchip manufactures.

MOLECULAR AND QUANTUM COMPUTING

Alternative Arcitectures for Nanocomputing

In addition to single electron transistors, two promising alternatives to traditional computers are molecular computing and quantum computing. These two methods are intimately related, yet deal with information on two different levels. Much progress has been made in these areas during the last years and both have been shown to be feasible replacements for semiconductor chips.

Quantum computing seeks to write, process and read information on the quantum level. It is at the nanoscale that quantum mechanical effects such as (the wave particle duality) begin to become apparent. Numerous scientists are seeking ways to store information within the quantum mechanical realm. This is not a simple task because of the delicate nature of quantum mechanical systems.

However, since the laws of quantum mechanics involves unintuitive principles such as superposition and entanglement, a quantum computer would be able to violate some rules that limit our classical computers. For instance, taking advantage of superposition would mean that a quantum bit of information, termed a qubit would be able to be used in several computations at the same time.

Taking advantage of entanglement would mean that the information could be processed over long distances without the classical requirement of wires.

Molecular computation is another method complimentary to quantum computing that seeks to write, process and read information within single molecules. One molecule that has proved most promising for molecular computation is Deoxyribonucleic acid (DNA).

DNA is a long polymer made of 4 different nucleotides that can be represented by the letters A, T, C and G. The order or sequence of these nucleotides within DNA provides the information for making protein, the main components of the molecular scale machinery used by living organisms to carry out life sustaining functions.

Mathematicians have figured out numerous ways to use DNA an the various proteins that come with it to carry out numerical computations that are notoriously difficult for silicon computers, namely NP-complete problems. The advantage that molecular computing using DNA has over conventional computing is that it is massively parallel. This means that each DNA molecule can function as a single processor, which greatly improves the speed of computation for complex problems.

MEDICAL APPLICATIONS

The other field in which molecular scale manipulation of matter is receiving abundant attention is medicine. Since all living organisms are composed of molecules, molecular biology has become the primary focus of biotechnology.

Countless diseases have been cured by our ability to synthesize small molecules commonly know as 'drugs' that interact with the protein molecules that make up the molecular machinery that keeps us alive.

Our understanding of how proteins interact with DNA, phospholipids and other biological molecules is what allows such progress. Living systems are able to live because of the vast amount of highly ordered molecular machinery from which they are built. The central dogma of molecular biology states that the information required to build a living cell or organism is stored in the DNA (which was described above for its use in molecular computation). This information is transferred from the DNA to the proteins by the processes called transcription and translation.

These processes are all executed by various biomolecular components, mostly protein and nucleic acids. Molecular biology is a field in which the study of these interactions has led to the discovery of numerous pharmaceuticals that have been enormously effective in curing disease.

Understanding of molecular mechanisms including substrate recognition, energy expenditure, electron transport, membrane activity and much more have greatly improved our medical technology.

So, what does this have to do with nanotechnology? First of all it shows the abilities of molecular scale machinery. Since the goal of nanotechnology is molecular and atomic precision, nanotechnology has much (if not everything) to learn from nature. Copying, borrowing and learning tricks from nature is one of the primary techniques used by nanotechnology and has been termed biomimetics. Secondly, our ability to design synthetic, semi-synthetic and natural molecular machinery gives us an enormous potential for curing disease and preserving life. An extensive textbook titled Nanomedicine has been written and does an excellent job of summarizing how nanotechnology is changing medicine.

NANOTECHNOLOGY FOR THE 21ST CENTURY

Experts in United States anticipate the world's population in 2050 to be approximately 8.7 billion persons. The world's population is growing, but its surface area is not. Compounding the effects of population growth is the fact that most of the earth's ideal farming land is already being utilized. To avoid damaging environmentally sensitive areas, such as rain forests, we need to increase crop yields for land currently in use. By increasing crop yields, through the use of biotechnology the constant need to clear more land for growing food is reduced.

Countries in Asia, Africa, and elsewhere are grappling with how to continue feeding a growing population. They are also trying to benefit more from their existing resources. Biotechnology holds the key to increasing the yield of staple crops by allowing farmers to reap bigger harvests from currently cultivated land, while preserving the land's ability to support continued farming. Malnutrition in underdeveloped countries is also being combated with biotechnology. The Rockefeller Foundation is sponsoring research on "golden rice", a crop designed to improve nutrition in the developing world. Rice breeders are using biotechnology to build Vitamin A into the rice. Vitamin A deficiency is a common problem in poor countries.

A second phase of the project will increase the iron content in rice to combat anemia, which is widespread problem among women and children in underdeveloped countries. Golden rice, expected to be for sale in Asia in less than five years, will offer dramatic improvements in nutrition and health for millions of people, with little additional costs to consumers.

Similar initiatives using genetic manipulation are aimed at making crops more productive by reducing their dependence on pesticides, fertilizers and irrigation, or by increasing their resistance to plant diseases. Increased crop yield, greater flexibility in growing environments, less use of chemical pesticides and improved nutritional content make agricultural biotechnology, quite literally, the future of the world's food supply.

CONCERNS

As biotechnology has become widely used, questions and concerns have also been raised. The most vocal opposition has come from European countries. One of the main areas of concern is the safety of genetically engineered food. In assessing the benefits and risks involved in the use of modern biotechnology, there are a series of issues to be addressed so that informed decisions can be made. In making value judgements about risks and benefits in the use of biotechnology, it is important to distinguish between technology-inherent risks and technology-transcending risks. The former includes assessing any risks associated with food safety and the behaviour of a biotechnology-based product in the environment.

The latter involve the political and social context in which the technology is used, including how these uses may benefit or harm the interests of different groups in society. The health effects of foods grown from genetically engineered crop depend on the composition of the food itself. Any new product may have either beneficial or occasional harmful effects on human health. For example, a biotech-derived food with a higher content of digestible iron is likely to have a positive effect if consumed by iron-deficient individuals. Alternatively, the transfer of genes from one species to another may also transfer the risk for exposure to allergens. These risks are systematically evaluated by FDA and identified prior to commercialization.

Individuals allergic to certain nuts, for example, need to know if genes conveying this trait are transferred to other foods such as soybeans. Labeling would be required if such crops were available to consumers. Among the potential ecological risks identified are increased weediness, due to cross-pollination from genetically modified crops spreads to other plants in nearby fields. This may allow the spread of traits such as herbicide-resistance to non-target plants that could potentially develop into weeds. This ecological risk is assessed when deciding if a plant with a given trait should be released into a particular environment, and if so, under what conditions.

Other potential ecological risks stem from the use of genetically modified corn and cotton with insecticidal genes from Bacillus thuringiensis (Bt genes). This may lead to the development of resistance to Bt in insect populations exposed to the biotech-derived crop. There also may be risks to non-target species, such as birds and butterflies, from the plants with Bt genes. The monitoring of these effects of new crops in the environment and implementation of effective risk management approaches is an essential component of further research. It is also important to keep all risks in perspective by comparing the products of biotechnology and conventional agriculture. The reduction of biodiversity would represent a technology-transcending risk. Reduced biological diversity due to destruction of tropical forests, conversion of land to agriculture, overfishing, and the other practices to feed a growing world population is a significant loss far more than any potential loss of biodiversity due to biotech-derived crop varieties. Improved governance and international support are necessary to limit loss of biodiversity.

What we know from our understanding of science and more than a decade of experience with biotech-derived plants is the following): There is no evidence that genetic transfers between unrelated organisms pose human health concerns that are different from those encountered with any new plant or animal variety. The risks associated with biotechnology are the same as those associated with plants and microbes developed by conventional methods.

SURVEY RESEARCH

Survey research over the past decade shows that biotechnology is not likely to become an important issue for most American consumers. Consumers find biotechnology acceptable when they believe it offers benefits and it is safe. Surveys have consistently found that a majority of American consumers are willing to buy insect-protected food crops developed through biotechnology that use fewer chemical pesticides, as well as more nutritious foods. American consumers also appreciate the role that biotechnology can play in feeding the world. Research shows that European consumers are much less supportive of all biotechnology applications.

Surveys since 1992 show that relatively few U.S. consumers have heard or read much about biotechnology. News about the cloned sheep pushed awareness to 50 per cent in March 1997. Surveys in the first three months of 2000 show that awareness has fallen back to just over one-third in the United States. Such trends reflect the fact that most people get their information about biotechnology from the media. Unfortunately, many consumers also do not understand some fundamental principles of biology. European consumer awareness is somewhat higher, but knowledge is still low.

Media coverage in the United States has generally been balanced (which helps account for our relatively high levels of acceptance). This is in sharp contrast to the European media, which have played upon fear of the unknown. The European media have also tended to accept opponents' claims without question. Another issue is that many people no longer have a connection to agriculture. In fact, research has shown that many consumers are unaware that all foods are derived from plants or animals that already have been genetically modified through traditional (but imprecise) breeding methods.

American consumers look to health professionals and scientific experts for credible information, but place relatively little trust in the activists who oppose biotechnology. Research shows that acceptance increases significantly when American consumers learn that organizations such as the National Academy of Sciences and the U.S. Food and Drug Administration have determined that biotech-derived foods are safe.

In contrast, European consumers express the most trust in those groups that oppose biotechnology. They have much less confidence in government, industry, or even scientists. American culture is more supportive and rewarding of new technology. Europeans tend to view food differently from U.S. consumers. In fact, some Europeans reject all American food products. Europeans also want to protect their small farms to maintain open space and rural employment. Such forces underlie much of the European anxiety about agricultural biotech - especially since it is seen as an "American invention."

Most of the industry leaders interviewed are quite enthusiastic about the benefits of biotechnology -- especially in terms of increased food availability, enhanced nutrition, and environmental protection. Most feel that biotechnology has already provided benefits to consumers. Almost all recognize that foods developed through biotechnology have already been part of consumers' everyday diet. They clearly do not agree with most of the opponents' claims and tend to have almost no trust in such groups. Their main concerns involve lack of consumer acceptance -- not the safety of the foods. They express high levels of confidence in the science and the regulatory process. In fact, almost none feel that biotechnology should not be used because of uncertain, potential risks. Most food industry leaders do not feel it is necessary to have special labels on biotech-derived foods. They express concerns that such labels would be perceived as a warning by consumers.

They also worried that the need to segregate commodities would pose financial and logistical burdens on everyone in the system - including consumers. Food industry leaders recognize a major need to educate the public about biotechnology. They look to third parties, such as university and government scientists to provide such leadership. Research shows that consumers will accept biotech foods if they see a benefit to themselves or society and if the price is right. Their responses to foods developed through biotechnology are basically the same as for any other food - taste, nutrition, price, safety and convenience are the major factors that influence our decisions about which foods to eat. How seeds and food ingredients are developed will only be relevant for a relatively small group of concerned, consumers.

The potential for public concerns has led several food companies to change their products to avoid biotech-derived ingredients. For example, Gerber Foods received threats from Greenpeace because they had determined the company was using biotech-derived food ingredients (mainly soy). The company firmly believes that the biotech foods are safe to consume.

Gerber agreed to drop some of its existing corn and soybean suppliers in favour of ones that can produce crops that are not genetically altered. It became an issue that is suddenly confronting other food companies. A private manufacturer in California, called Healthy Times Natural Food has switched from Canola oil (which sometimes is genetically modified) to safflower oil after facing questions from Greenpeace. The controversy is due in part to the fact that the organic industry is using public concern as a tool for marketing their products as free of biotech ingredients.

NATIONAL AND INTERNATIONAL POLICY

National governments and international policy making bodies rely on food scientists and others to develop innovations that will create marketable food products and increase food supplies. Governments also rely on scientific research because they are responsible for setting health and safety standards regarding new developments.

International organizations can suggest policy approaches and help develop international treaties that are ratified by national governments. Economic success in the competitive international market demands that food production become more efficient and profitable. National governments and international organizations support food biotechnology as a means to avoid global food shortages.

Many policy making bodies are also trying to balance support of the food biotechnology industry with public calls for their regulation. Such regulations are necessary to protect public health and safety, to promote international trade, conserve natural resources, and account for ethical issues. The majority of processed foods on the market contain soy or corn ingredients that come from GM plants. To date none have posed a food safety risk. The chief safety concerns are the potential to alter nutrient content or introduce allergens. Federal agencies involved in biotechnology regulation include the U.S. Department of Agriculture (USDA) which evaluates agricultural production processes for all foods; the Food and Drug Administration (FDA), which evaluates whole non-animal foods (seafood), food ingredients, and food additives; and the Environmental Protection Agency (EPA), which evaluates plants with insecticidal properties.

Developers of GM plants and biotech-derived foods are required to consult with FDA prior to the commercialization of the product. This consultation procedure entails a science-based safety assessment of the product that focuses on the protection of the consumer, developer, and the environment. Thus developers, have a strong incentive to cooperate fully with FDA and the other agencies prior to marketing their products.

LABELING ISSUES

Labeling food derived from GM plants and animals is an important, but complex issue. Some consumers and consumer groups believe they have a right to know whether biotechnology was used to produce food. Others believe labeling is not necessary if foods are essentially equivalent in composition. Food labels are regulated by FDA and, in some cases, by USDA.

Regulatory agencies are concerned with ensuring that food labels are both true and not misleading. In accordance with their statutory mandate, the FDA has determined that a food product should be labeled as a product of biotechnology only if it has been changed in some significant way. The U.S. Court of Appeals recently reaffirmed that the FDA's labeling policy is correct. Based on the input from three public hearings, the FDA has recognized that the most effective way to allow for informed choice will be a system of voluntary labeling for foods not produced through biotechnology.

If the demand for "GM-free" food is real, a market will become viable. In this case, meaningful choice can be provided to concerned consumers without imposing costs on or denying benefits to the majority of consumers, who support biotechnology. This may be the only viable and cost-effective option for the food industry (given the costs and difficulties associated with segregation and identity preservation.)

National surveys conducted with American consumers between 1997 and 2000 found about three-quarters of consumers supported this FDA labeling policy. There is evidence from focus groups that US consumers are already overwhelmed by the level of detail on food labels and mainly look for relevant information about nutrition. Recent focus groups have demonstrated that the wording on labels has a significant effect on consumer understanding and acceptance of biotechnology.

Any label messages need to be simple and clear. However, labeling will be counterproductive without education. It is important to understand when FDA will and won't require a product label. The same labeling laws that apply to all other foods and food ingredients apply to products of food biotechnology.

Requirements for all food labels mandate proper identification of products and notice of health or safety concerns. Labeling would be required in the following instances - but not simply because the products were made using biotechnology.

- Potential food allergy is an example of a health or safety risk that would mandate a product label (*e.g.*, if a gene has been transferred from a known allergen, such as a peanut).
- When substantial changes have been made to the expected nutritional composition of a food (*e.g.*, vitamins have been added to a particular type of produce).
- If the common identity of a food product has been changed (*e.g.*, if the composition of a fruit or vegetable becomes significantly different it would have to have a different name).

Labeling of biotech-derived foods will be required in some countries (such as Japan and the European Union) for cultural and religious reasons. These countries also believe labeling should be required simply because the consumers want to know more about the content of the food or how it was produced. Some have suggested that such labeling initiatives have been driven by economic and political interests rather than by scientific need. Consumers in these countries will ultimately end up paying even higher prices for their food as a result of such strict labeling policies.

The applications of biotechnology are so broad, and the advantages so compelling, that virtually every industry is using this technology. Developments are underway in areas as diverse as pharmaceuticals, diagnostics, textiles, aquaculture, forestry, chemicals, household products, environmental cleanup, food processing and forensics to name a few.

Biotechnology is enabling these industries to make new or better products, often with greater speed, efficiency and flexibility. Biotechnology holds significant promise to the future but certain amount of risk is associated with any area. Biotechnology must continue to be carefully regulated so that the maximum benefits are received with the least risk. Biotechnology is at a crossroads in terms of public acceptance. Many Americans have not yet formed a solid opinion on this complex issue. International developments over the next few years will certainly have a major influence on the long-term viability of biotechnology. The future of the world food supply depends upon how well scientists, government, and the food industry are able to communicate with consumers about the benefits and safety of the technology.

Several major initiatives are under way to strengthen the regulatory process and to communicate more effectively with consumers. Both the USDA and FDA have opened their regulatory systems to outside review and public comment. The biotechnology industry, university scientists and others are also conducting educational programmes.

These should further strengthen consumer confidence. This partnership among the public and private sectors will support these emerging technologies that will prove vital to the U.S. economy and the developing world in the new millennium. Even Europe will soon find the real benefits of biotechnology compelling.

2

Nanoparticles

The rapidly developing nanoscience and nanotechnology reveal many exciting features of materials at the nanometer (10^{-9} meter) scale, and the possibility of manipulation of such features to create novel materials or products that are previously unthinkable. Many believe that nanoscience and nanotechnology development will become a new industrial revolution.

On the other hand, together with the numerous beneficial potentials, scientists also warn against possible adverse health and environmental impacts brought by the vastly unknown physicochemical properties of materials at the nanoscale. Among the different environment, health and safety issues, nanomaterials, in particular, nanoparticles probably cause the highest concern because of their potential of rapid uptake by biological system, unknown distribution pathways and potential interactions with various components of the biological system.

OBSERVATION

Nanoparticles are previously known in the air pollution field as ultrafine particles, which are defined as particles less than 100 nm in aerodynamic equivalent diameter (AED). Some ultrafine particles are generated by natural processes such as volcanic activities and wild fires. Others are derived incidentally from anthropogenic sources, *e.g.* engines and incinerators. In general, high energy activities involving organic or inorganic all have the potentials of producing ultrafine particles as one of the byproducts. On the other hand, with the development of nanoscience and nanotechnology, many nanoparticles are purposely made for particular functions. These are usually called engineered nanoparticles, and are defined as engineered particles with at least one dimension smaller than 100 nm.

Nanoparticles have unique physicochemical properties that are not found in their parent materials. In general, they have much higher reactivity, and because of their ultra small size, they can easily penetrate skin or cells, rapidly distribute in human body, and even directly interact with organelles within cells. Their huge surface area to mass ratio increases the chemical activities and therefore allows them to become efficient catalysts. These increased chemical and biological activities result in many engineered nanoparticles being designed for specific purposes, including diagnostic or therapeutic medical uses and environmental remediation that require introduction of these novel materials into human body or the environment.

The novel properties of nanomaterials, however, are a two-sided sword. The same properties that allow the wonderful beneficial uses also imply potentially devastating adverse health effects. Rapid uptake through skin and epithelial cells, capability to translocate along neurons, plus the known toxicity of some parent materials warrant careful environment, health and safety evaluation side by side with every nanomaterial application study.

NANOMATERIAL TOXICITY

Researchers worldwide are only beginning to look into toxicity of various nanomaterials. Knowledge of this newly emerged field, sometimes called "Nanotoxicity", is continuously being generated. The following is a discussion of certain known toxicity of several common groups of nanomaterials.

Carbonaceous nanomaterials-Fullerenes, consist of repeating hexagonal and pentagonal rings of carbon atoms. They include buckyballs (C_{60}) and other spherical structures, and single wall and multi-wall carbon nanotubes. Buckyballs are extremely stable. They can trap other atom or molecule inside the carbon structure. Possible applications include superconducting material, drug delivery system, lubricant and catalyst. Carbon nanotubes can also be superconducting, and can be made into mechanically super-strong materials which are 100 times the strength, but one sixth the weight of steel, among many other potential applications.

Carbon has always been considered a non-toxic element, and not very water soluble. However, recent studies found that buckyballs have considerable water solubility, and can cause lipid peroxidation in fish brain after 48 hours at a concentration of 500 parts per billion (ppb). This kind of oxidative stress is usually a first sign of biological damage. Single-wall nanotube has been found to cause other oxidative stresses such as free radical formation and depletion of antioxidants in cells. Metal impurity in synthesis of nanotube may also add to the toxicity. Besides, physical dimensions of some nanotubes may cause pulmonary fibrosis and cancer similar to the effect of asbestos.

Another example of potential toxicity is quantum dots (Q-Dots), which are nanoparticles of semiconducting materials. Their sizes are so small that they interfere with quantum behaviour of atoms. One of the results is they display tunable, based on particle size, absorption and emission spectrum over visible wavelength. This property gives quantum dots a strong potential in medical imaging. However, the current quantum dots are made of a cadmium-selenium (CdSe) core with a zinc sulfide (ZnS) shell, and all three metals are known to be toxic to different extents. The high surface area and rapid distribution of the ultra small particles are expected to further escalate the potential toxic effects once introduced into a biological system.

Some nanomaterials have already found their ways into commercial products, such as titanium oxide (TiO_2) for photochemical oxidation, self-cleaning paints, sunscreen products, etc.; and zinc oxide (ZnO) in cosmetics and sunscreen products. Other nanoparticles are being used in specialized applications, *e.g.* iron particles for oxidation of contaminants in groundwater, copper oxide (CuO) for antimicrobial activities; and iron oxide (Fe_2O_3) as a contrast agent for magnetic resonance imaging. Although most of these parent materials are not considered to be significantly toxic, the ultra small size of the particles again may create unknown toxicity or environmental impacts. There were already studies that showed the toxicity of nanosize TiO_2increases proportionally to the surface area, not the mass as in conventional toxicology. A recent study also measured free nanosize TiO_2 particles off the surface applied with a paint containing the nanoparticles.

OCCUPATIONAL HAZARDS OF NANOMATERIALS

The environmental, health and safety concerns of nanomaterials in fact span from research safety, manufacturing safety, to consumer (product) safety, and include environmental impacts throughout the life cycle, until final disposal of the nanomaterial products and wastes. Among these concerns, research and pilot scale production safety is most relevant to the research and development work at HKUST. Due to the rapid development and many uncharted areas in nanoscience and nanotechnology, the practice of occupational health and safety also faces many challenges. As for many conventional workplace contaminants, airborne route is considered the most crucial for worker protection.

The shape and size of airborne nanoparticles, and the conglomeration status are some of the main factors determining the hazard. Nanoparticles in liquid phase may present skin hazard when contacted. The hazard of the liquid matrix, its potential interaction with the nanoparticles, *e.g.* serving as a carrier that facilitates skin penetration, and the conglomeration status of the particles will affect the overall hazard. In either the airborne or liquid phase, the nanomatierals have the potential to adversely impact the environment if they are not properly contained during usage and disposed of afterwards.

CHALLENGES IN ASSESSING AND CONTROLLING

A major challenge in evaluating hazard posed by airborne nanoparticles is to find a (set of) suitable exposure metric(s). It will be necessary to consider number concentration and surface area, other than the conventional mass concentration, then appropriate sampling and analytical methods, and eventually health-based standards, will have to be developed. Quite a few number concentration direct reading instruments originally designed for ultrafine particle monitoring are available, but their applicability to monitoring of nanoparticles in a work environment remains to be tested. Also the reliability of the detection technology and how "direct" the reading really is are also main concerns to be addressed. In order to differentiate engineered nanoparticles of interest with natural or incidental nanosized particles that may be present in the background, examination by transmission electron microscopy (TEM) remains the golden standard for characterizing and enumerating nanoparticles.

At least one passive sampling method coupled with enumeration by electron microscope has been developed, its precision and accuracy are yet to be determined. Governmental agencies of developed countries such as the National Institute for Occupational Safety and Health (NIOSH) of the US, and Health and Safety Executive (HSE) of the UK, are actively working to fill up these knowledge gaps in assessing hazards of airborne nanoparticles.

Another major challenge is control technologies. Traditionally, high efficiency particulate air (HEPA) filters are used for removing airborne particulates. However, before the rise of nanomaterial research, it has never been proved whether HEPA filter can capture nano scale particles. Theoretically the filtering efficiency of HEPA increases with decreasing size below 0.1-0.3 μm, which is the most difficult size range to capture according to physics principles of filtration by fibrous materials. Previously there were only measured data points down to 0.01 μm (10 nm), beyond which the filtration efficiency was by extrapolation.

However, according to preliminary data presented in the Second International Symposium on Nanotechnology and Occupational Health (Minneapolis, October 2005) by a group of researchers from University of Minnesota, 3M and NIOSH, HEPA filter is capable of removing more than 99.97 per cent of particles down to the range of 2-4 nm. No significant thermal rebound was observed. This means with our current knowledge, HEPA filter can be deployed as an effective means to contain and control airborne nanoparticles at least down to the low nm size range.

As for skin protection, most of the glove and protective clothing materials currently in use have not been tested against nanosize particles. Presumably the materials should be protective against nanoparticle conglomerates. Some suggested treating airborne nanoparticles as toxic gas in terms of personnel protection. In any case, until information on interaction of nanoparticles and protective material is available, skin protection should at least be adequate to prevent penetration of the carrier medium of nanoparticles in the liquid phase.

Either in air or liquid phase, nanoparticles tend to agglomerate instead of existing in single particle form. This is helpful from a hazard control point of view. However, certain future applications or manufacturing processes may rely on dispersed nanoparticles. Therefore it is prudent to plan safety measures assuming the worst case scenario. Some basic precautionary measures have been suggested by organizations such as HSE of UK and NIOSH of US. These include treating all nanomaterials that may possess unique physical chemical properties as "new chemicals" for regulatory purpose.

Material and safety information raw material can only be viewed, at best, as a starting point or reference for assessing the risk of the nanomaterial product. A good example is carbon, which is practically non-toxic as a parent material; but a group of scientists have already established a toxicity level of oral lethal dose 50 (LD50) of 300 mg/kg for C_{60} buckyballs, which is considered moderately toxic.

An important concept that worth mentioning is "Responsible Nanotechnology", which basically says, there are many promising applications for nanoscience and nanotechnology, however, there are also uncertainties in potential health and environmental effects. Therefore we should not repeat the errors we made before in cases like DDT and chlorofluorocarbons, where people focused on the beneficial use of the new chemicals so much that they did not carefully check for potential health and environmental impacts until damages were done. Concurrent development of technology and environment, health and safety assessment and controls must go hand in hand in this potentially revolutionary field of research.

The UK Royal Society and Royal Academy of Engineering published a report titled "Nanoscience and nanotechnologies: opportunities and uncertainties" in July 2004.

They proposed the following general guidelines, which form a good foundation for responsible nanotechnology development:

- Treat nanomaterials as if they are hazardous
- Seek to remove them from waste stream
- Don't apply free nanomaterials to environment unless benefit clearly outweigh risk
- Assess potential impact throughout lifecycle of nanomaterial

In early 2006, the Government supported the establishment of Nanotechnology and Advanced Material Incorporated (NAMI) under the direction of HKUST to facilitate nanoresearch in Hong Kong and transfer of nanotechnology to local industries.

This undoubtedly elevates HKUST to a leading position in nanotechnology development in Hong Kong. It also brings into focus the need to integrate EHS assessment into nanotechnology development all the way from research, pilot production, manufacturing, consumer use and final disposal. This presents a great opportunity for us to realise the idea of "Responsible Nanotechnology", *i.e.* to not only take advantage of the benefits that nanotechnology brings us, but also ensure that we do not enjoy these benefits in the expense of health, safety or the environment.

HKUST

Nanomaterial and nanotechnology is a rapidly evolving field, and is one of the focus areas of research at HKUST. The environmental health and safety knowledge about nanomaterials and nanotechnology is also being generated by many researchers worldwide. Even though there are considerable uncertainties, it is prudent to take precautionary measures to minimize the potential of adverse health, safety, and environmental impacts. The following are elements of the Nanosafety Programme at HKUST. Due to rapid development of this field, these programme elements will be revised when necessary.

- *Facility Review*: HSEO has been involved in the planning of laboratory facility for the Institute of Nanomaterials and Nanotechnology (INMT). When faculty members establish other research facilities for nanomaterials and nanotechnology, they should involve HSEO at the planning stage to ensure necessary safety installations are included in the physical facility, safe operational protocols are developed, and suitable personal protective equipment is prescribed and made available to lab workers.
- *Research Proposal Review*: A nanosafety section has been added to the research proposal safety review form under the Committee of Research Practices (CRP), where principal investigators are requested to declare whether any nanomaterial will be used, and if so, the physical and chemical properties of the nanomaterial, known toxicity, use conditions, proposed control measures, and final disposal arrangement. Based on the information provided, the Safety Panel of CRP will suggest necessary precautionary measures.
- *Laboratory Visit*: At this early stage of development, one of the crucial elements is to establish a database of nanoresearch activities on campus. Based on the information collected from the research proposal review, HSEO staff will visit research groups working on nanoscience and nanotechnology to collect information on the nanomaterials, work processes, potential hazards involved, and waste disposal practices. Since safety standard is yet to be developed, available guidelines such as those from NIOSH and HSE will form the basis of recommendations on EHS issues.
- *Safety Precautions and Safety Installations*: As mentioned in Section E above, HEPA filter is a proven method to contain and control airborne nanoparticles. Research should conduct all processes that may give rise to airborne nanoparticles inside one of the following safety devices:
  - Biological safety cabinet (BSC), if a dust-free environment is not critical, a Class 1 BSC can be used. Otherwise, a Class 2 BSC

can provide both a dust-free environment and protect lab workers at the same time. However, BSC cannot contain volatile or toxic chemicals, if these are to be used together with nanoparticles. One should use a fume hood equipped with HEPA filtration

- Fume hood equipped with HEPA filtration
- Glove box

Be reminded that BSC, fume hood and HEPA filter all need periodic testing to ensure they are properly functioning. Please contact either FMO-LS or HSEO if you have questions about such periodic testing.

- Safety Precautions and Personal Protective Equipment (PPE)
  Laboratory workers handling nanoparticles or other nanomaterials should follow these guidelines:
- Always ensure personal hygiene and observe good general laboratory practices, including no drinking and eating in the laboratory, always wearing lab coat and safety glasses, etc.
- Consider the hazards of the parent materials in assessing hazards of nanomaterials. However, remember the nano features of the materials most likely will present many other physicochemical and toxicological properties not found in the parent materials.
- Never handle nanomaterials with bare hands or expose such materials to bare skin. Until further information is available, use chemical resistant gloves and protective clothing. The material of these PPE items should at least be protective against the carrier medium, such as the solvent, containing the nanomaterials.
- If it is necessary to handle nanoparticles outside of a safety installations specified in the previous section, use appropriate PPE such as a properly fitted elastomeric respirator with HEPA filter cartridge. Please contact HSEO for requirements of a respirator user. If the exposure risk to nanoparticles is considered low, particulate respirators with a filtration efficiency equivalent to HEPA (*e.g.* the 100 series approved by NIOSH or P3 by EC standard) may be used.
- A vacuum cleaner equipped with a HEPA filter may be used to remove nanomaterials from reactors or work areas. However, the integrity of the HEPA filter must be periodically tested to ensure proper functioning.
- Dispersed airborne nanoparticles, like many other airborne fine particles, can become a fire and explosion hazard. Care must be taken to prevent ignition.
- Equipment in contact with nanomaterials should be evaluated for residual hazard before servicing, reuse for other purposes, or disposal.
- Contact HSEO for assistance in evaluating or controlling hazards related to nanomaterials.
- Waste Disposal

  - Waste containing nanomaterials should be disposed as chemical wastes as a minimum. See Chapter 16 for general chemical waste disposal procedures.
  - Researchers should seek to remove the nanomaterials from the waste stream where possible.
  - Alternatively, researchers should destruct the nano features of the materials where feasible. For examples, acid may be added to waste solution containing quantum dots to dissolve the nanoparticles before disposal.
  - Following the principle of "Responsible Nanotechnology, and the concept of minimizing potential environmental impact, researchers should develop suitable destruction method for nanomaterials as a part of their research.
  - Contact HSEO for assistance in disposal of nanomaterial waste.
- *Exposure Monitoring*: Even though standard method has not been established for measuring exposure to nanoparticles yet, some conventional exposure monitoring methods can be employed to provide an estimate of exposure. Some field instruments can also be used to measure airborne nanoparticles. HSEO will continue to monitor the development of exposure measuring technology for nanomaterials.
- *Continuous Improvement*: As the field is rapidly evolving, HSEO will continue to monitor the development, especially the environment, health and safety aspects, and will introduce necessary amendments to the nanosafety programme. We urge all researchers to integrate EHS considerations into their research projects, and contribute their expertise to the responsible development of nanotechnology at HKUST.

The continuing revolution in computer hardware is packing ever more logic gates in ever smaller volumes. Unfortunately this produces ever more heat, which will limit the feasible packing density and performance unless the energy dissipated by each logic operation can also be dramatically reduced.

To reduce the energy dissipation of a logic operation below ln(2) kT (near thermal noise) requires the use of reversible logic for fundamental thermodynamic reasons. Extrapolation of current trends suggest this limit will become significant within one to two decades. Many real devices can be viewed as electrically controlled switches, so a method of using an abstract switch in a reversible manner is useful. Two methods of using switches to implement reversible computations are discussed in this paper. The first method has an energy dissipation which is proportional to the square of the error in the voltage, while the second method has an energy dissipation which can in principle be reduced indefinitely by slowing the speed of computation.

The first method is basically an extension to "pass logic" which has been previously used with both nMOS (hot clock nMOS) and CMOS transmission gates to achieve low energy dissipation. The second method is a novel thermodynamically reversible logic system based on CCD-like operations which switches charge packets in a reversible fashion to achieve low energy dissipation. There is now a fairly extensive literature on reversible computation which shows that the energy dissipation per device operation cannot be reduced below ln(2) kT (where k is Boltzman's constant and T is the temperature; kT is roughly the thermal energy of a single atom) if the device is not reversible.

For the last 50 years the energy dissipation per gate operation has been declining with remarkable regularity. Extrapolation of this trend shows the energy dissipation per device operation reaching kT by the year 2015. (This assumes that T is 300 Kelvin-more on this later). To gain some perspective on this consider that an "AND" gate which has a power supply of one volt and which allows 100 electrons to go from that one volt supply to ground during the course of a switching operation will dissipate 100 electron volts. Although 100 electron volts is about 4,000 times kT (and well above the theoretical limit), it will be difficult for simple improvements of irreversible approaches to reach even this level of energy dissipation.

Extrapolating present trends, we should reach 4,000 kT before the year 2000, *e.g.*, within ten years. While reversible logic is an absolute necessity if we are to reduce energy dissipation per device operation below kT, it is a useful heuristic for designing systems that have low energy dissipation even when the actual energy dissipation is well above kT.

Even if we do develop irreversible devices that approach ln(2) kT, a computer operating at room temperature at a frequency of one gigahertz with 10 ^ 18 logic gates packed into a volume not much larger than a cubic centimeter would dissipate over three megawatts. The drive for ever greater computational power with ever more densely packed logic elements will eventually require that a single logic operation dissipate orders of magnitude energy less than kT.

If we are to realise the full potential of nanoelectronics and molecular logic devices, at some point in the future we will be forced to use thermodynamically reversible logic elements for a substantial fraction of logic operations.

We can state quite confidently that one of three things will occur:

- The historic rate of decreasing energy dissipation per device operation will slow or halt in the next one or two decades or
- We will operate computers at lower temperatures or
- We will develop new methods of computing in a reversible way that can beat the kT barrier.

The first option is unattractive. First, cooling is a major problem. The heroic cooling methods used in the Cray 3 supercomputer to remove the heat generated by the computer's operation suggest that failure to reduce energy dissipation per gate operation would be a major limiting factor in future computer performance.

While both air-cooling a single chip that dissipates 150 watts and water cooling of a chip dissipating 790 watts/cm ^ 2 have been demonstrated, and more effective cooling methods should be feasible in the future, we will eventually reach a limit to heat removal.

Second, in many applications power is limited. Portable computers, remote sensors and other isolated systems have only a limited amount of power available. Third, although the raw cost of electrical power is not yet a major limitation it would become so in the future if reductions in energy dissipation did not keep pace with advances in other areas.

The Wall Street Journal said "Computer systems currently account for 5 per cent of commercial electricity use in the U.S., with the potential to grow to 10 per cent by the year 2000." Operating computers at a lower temperature will not reduce overall energy dissipation. If we operated future devices at 3 Kelvins we would reduce kT by a factor of 100 and so could reduce energy dissipation per logic operation by a similar factor-but for fundamental thermodynamic reasons the coefficient of performance of the refrigerator for the system can be at best 3K/(300K–3K) or about 0.01.

Thus, the lower energy required per gate operation will be almost exactly balanced by the increased energy needed by the refrigerator. In many applications refrigeration is not an attractive option. Lap top and portable computers, embedded systems, various "smart" appliances and other applications make the use of refrigeration undesirable. Factors other than net energy savings can make low temperature operation worthwhile. For example, some potentially attractive devices don't operate at higher temperatures.

In large computers operating in stable environments (the traditional computer centre, for example) refrigeration might be attractive, particularly if it permits the use of devices that provide much better performance but which require low temperature for their operation (Josephson junction devices might be an example). Refrigeration *per se*, however, does not seem too attractive. Finally, and most attractively, we could use reversible computation to reduce energy dissipation. This would, in principle, allow energy dissipations indefinitely below kT per logic operation.

While some barrier should eventually be encountered, the use of reversible logic should allow us to continue current trends in energy dissipation per logic operation for the longest possible time. It is interesting to note that certain uses of pass logic, transmission gates, and hot clock nMOS do in fact perform some logic operations in a thermodynamically reversible fashion.

Research in the design and utilization of reversible logic to futher lower energy dissipation will simply recognize and make explicit a trend that has already begun. Further research in this area is the most appropriate response to the rather limited range of possibilities that face us.

REVERSIBLE ARCHITECTURES

One concern about this approach is the need to use reversible computer architectures. Such architectures are entirely feasible. A wide variety of computations can be done in a logically reversible manner. Bennett concluded: "...for any e > 0, ordinary multitape Turing machines using time T and space S can be simulated by reversible ones using time O(T ^ (1+e)) and space O(S log T) or in linear time and space O(S T ^ e)." This general result shows that even arbitrary irreversible computations can be mapped into reversible computations.

Specific irreversible computations can often be mapped into reversible computations with little or no loss of efficiency. Even if we do not adopt new reversible computer architectures, simple applications of reversible computation could be made within the framework of existing architectures.

A typical computer executes a sequence of instructions, and each instruction will typically change the contents of a single register or memory location. Although loading the result of the instruction into a register will normally be irreversible (it erases the previous contents of the register) it is still the case that all other operations performed by the computer during instruction execution could in principle be made reversible.

Thus, although we would have to dissipate roughly kT energy for each bit in the output register for each instruction execution, we need not use dissipative logic throughout the computer (as is currently done).We can go one step further without making significant changes in computer architecture: the simple register-register add instruction $R_1 = R_1 + R_2$ is logically reversible. With proper hardware, this particular instruction could in principle be made to dissipate as little energy as we desired. While non-reversible instructions (*e.g.*, R1 = 0) would still dissipate greater energy, the overall energy dissipation of the computer could be reduced significantly if a reversible approach were developed and used to implement the reversible instructions. Of course, once the energy-wasting instructions were identified, compilers would learn to avoid using them. Today, compilers can optimize for speed of computation and for the amount of memory used.

In the future, compilers could also optimize a programme to minimize the energy dissipated during computation, *e.g.*, to minimize the number of bits that are erased. This would provide an entirely evolutionary path from the current irreversible designs to computer architectures that were as reversible as was practically feasible. While it is perhaps not obvious how far this trend can go it is clear that a very large percentage of computer operations can be made reversible-perhaps a remarkably large percentage.

REVERSIBLE DEVICES

Several reversible devices have been proposed over the years. Von Neumann and independently Goto proposed the parametron which encodes information in the phase of an oscillation. Fredkin and Toffoli proposed an electronic reversible logic family which used switches, capacitors and inductors. By turning the switches on and off at just the right times, charge could be transferred between the capacitors through the inductors with minimal energy loss.

By increasing the size of the inductors it would be possible to create a family of ever slower but ever more energy efficient circuits. As the LC time constant is increased, the energy dissipation can be decreased to an arbitrary extent.

A proposal by Likharev based on Josephson junctions "...is particularly significant, because it is a genuine example... of a system that has frictional forces proportional to velocity" according to Landauer. The parametric quantron is based on Josephson junctions and operates at low temperatures. It should dissipate less than kT when operating at 4 kelvins and with a switching time of 1 nanosecond. Modern "high temperature" superconductors might allow operation at relatively high temperatures, including that of liquid nitrogen.

Even these temperatures would still be a significant disadvantage. Molecular-scale mechanical reversible "rod logic" has been proposed by Drexler which should dissipate less than kT when operating at room temperature at a speed of 50 to 100 nanoseconds. Drexler's proposal is also significantly smaller than the parametric quantron. While the ability to fabricate Drexler's proposal is some decades away, it provides a good argument that molecular-scale reversible devices should be feasible.

Although the parametric quantron can be used in a reversible fashion to achieve low energy dissipation it uses magnetic fields to couple logic devices, requires low temperatures to operate and will likely prove resistant to scaling to the molecular size range. By contrast, Drexler's mechanical proposal might well prove smaller than even the smallest of future electronic devices. Mechanical devices depend fundamentally on the position of the nuclei of the atoms of which they are made. The position of the nucleus of an atom can be much more precisely known than the position of an electron: quantum uncertainty in the position of the electron is much greater because the mass of the electron is much smaller.

If device function depends on the position of an electron then device size will be larger than if device function depends on the position of the nucleus. Such molecular mechanical proposals will, however, almost certainly be slower because of the greater mass of the nuclei of atoms as compared with the mass of electrons. In this paper we discuss simple methods of using relatively conventional electronic switching devices in a reversible fashion which can be scaled to a very small size. The methods described here allow a simple voltage-controlled switch to be used in a reversible manner, both to implement combinational logic and also to iterate an arbitrary reversible logic function.

Other methods are feasible. For example, helical logic—an elegant form of reversible electronic logic—has recently been proposed by Merkle and Drexler. Voltage controlled switches of various types have long been used in electronic devices. Not all voltage controlled switches are suitable for use in low energy designs because of their intrinsic energy dissipation.

Traditional relays actuated by electromagnets dissipate comparatively large amounts of energy. Even if a switch were suited for low energy operation it would still dissipate energy if it were turned on when there was a voltage across it. The abrupt rush of current that normally occurs when a switch is turned on would insure this. Most previous logic designs based on switches operated in this manner (though the proposal by Fredkin and Toffoli avoids this problem, and the proposal by Seitz et. al. eliminates dissipative switch closures for combinational logic, although not for sequential logic).

The switch is turned on for the purpose of draining charge when the charge is no longer wanted. That is, a logic "1" might be encoded by a charge on a capacitor. Turning on a switch to drain away that charge would dissipate energy even if the switch itself were perfectly non-dissipative. A related but distinct line of research also emphasizes the importance of energy dissipation, but assumes that the losses involved in dissipatively switching a wire from one logic state to another are fixed and cannot be altered, *e.g.*, that slowing the switching speed will not reduce energy dissipation.

Even assuming that this is true, clever design of the circuit can greatly reduce the number of times it is necessary to take such a dissipative action. Further research which combines both this approach and the approaches described here would seem fruitful. Even relatively recently some authors have argued that switching devices must fundamentally dissipate more than kT regardless of how slowly they operate. It is worth noting, therefore, that charging and discharging a condenser need not dissipate the energy involved.

For example, a simple LC oscillator will repeatedly charge and discharge a condenser. Although the energy stored on the condenser is 1/2 CV ^ 2, this energy need not be dissipated during a charge/discharge cycle.

If the Q of the circuit is high the energy dissipation per cycle will be small. In addition, the energy dissipation of such an oscillator depends on frequency. By using larger condensers and inductors the frequency of operation can be slowed and the fraction of energy dissipated per cycle reduced. If all linear dimensions of a device are scaled in size in proportion to some characteristic length L then capacitance and inductance are proportional to L while frequency and resistance are proportional to 1/L. The Q of the circuit is proportional to L so the percentage of the total energy in the oscillator that is dissipated per cycle can be reduced simply by increasing all linear dimensions equally. Different scaling methods can produce better results. If we assume that a fixed capacitive load is charged through a fixed resistance by a voltage that is ramped up over a period of time T, then the energy dissipated by the resistance is approximately 1/2 CV ^ 2 × 2RC/T (for RC << T).

This dissipation can be reduced as much as we might want by increasing T (and leaving the circuit components unchanged). Koller and Athas have called this adiabatic charging. We will call it "thermodynamically reversible" or simply "reversible," relying on context to make it clear that we are referring to an asymptotically non-dissipative process. Note that the term "reversible" can also be used to mean "logically reversible" with no implication that it is asymptotically non-dissipative.

A logically reversible process might be highly dissipative, so it is sometimes necessary to distinguish between "logically reversible" and "thermodynamically reversible. The situation is quite different for a step function in which the voltage is ramped up instantly (T=0). This process dissipates 1/2 CV ^ 2. Charge falls through the fixed potential drop in a dissipative fashion much as water goes over Niagra falls. Although this is the method in common use today to charge and discharge the various capacitive loads in computer circuits, it is not the method that we will be considering.

By using reversible charging, further increases in the switching time can be used to decrease the energy dissipation per logic operation. This, of course, results in a slower speed of operation leading to a trade-off between speed and energy dissipation. How to arrange matters so that current flows smoothly and reversibly is the principle topic of the rest of the paper.

This is challenging because the basic switching devices are:

- Abrupt and discontinuous and
- Pass through a dissipative intermediate state (which is neither fully on nor fully off) while switching.

THE FIRST METHOD

The first method uses a conventional voltage controlled switch but in a somewhat unusual manner. In essence we never turn a switch on or off when there is either a voltage across it or current going through it. This mode of operation is reversible. We extend this approach to implement a sequential circuit which can iterate the computation of a reversible logic function. It is also related to "dry switching," a method for using relays which avoids turning relays on or off when there is a voltage across them. The primary purpose of dry switching is to avoid deterioration of the relay contacts, rather than minimize switching energy.

We also adopt the following convention for turning a switch on and off: when the control input to the switch is at logic "0" (low) the switch is turned on and current can flow from the input to the output or from the output to the input. When the control input is "1" (high) the switch is turned off and current cannot flow. The reader should note that this convention is not the usual one. We adopt it instead of its opposite —where a "1" turns the switch on and a "0" turns it off— because it is more intuitive when applied to the second proposal discussed later in this paper, the CCD-based reversible logic element.

The proposed mode of operation is fundamentally clocked and multi-staged. The outputs of stage i drive the inputs of stage i+1. For the moment, we assume the output of stage 0 is simply given. That is, a set of input lines are assumed to have fixed voltages upon them representing the input data.

The value of the succesive logic levels can then be computed in turn. The logic values (voltages) produced by stage 1 are computed from the logic values given by stage 0; The logic values produced by stage 2 are computed from the logic values produced by stage 1; etc. Simple pass logic can be employed to implement combinational logic. A "NOR" operation in this logic is simply two switches in series. A clock signal (initially at logic 0) is connected to the input of the first switch.

If the control inputs of both switches have settled to logic 0, then both switches will conduct. The clock input connected to the first switch is slowly raised (from logic 0 to logic 1). Both switches will conduct and the output will gradually shift to a logic 1. If either input is 1, one of the switches will not conduct and the output will remain at 0.

MAKING AN N-LEVEL CIRCUIT

We assume that we have an N level combinational circuit, and that N "clock signals" are provided. Initially, clocks 1 through N are low (logic 0), and the outputs of stage 0 are at some preset values. The outputs from stage 1 are computed as simple logical functions from the outputs of stage 0. The outputs from stage i are used as the inputs to stage i+1. A "NOR" and "NAND" gate from stage 1. The NAND gate is logically complete so any arbitrary boolean function can be implemented by using an appropriately connected collection of them. The clock signals are now raised in turn.

The output of "stage 0" is assumed to be valid prior to the start of operations, and clock 0 is unspecified (for now). Clock 1 is raised, and the outputs from stage 1 become valid. Clock 2 is raised, and the outputs from stage 2 become valid. This sequence continues, until clock N is raised, and the outputs from stage N become valid. A single stage in the N level circuit could in principle implement an arbitrary boolean switching function by an appropriate arrangement of switches. This might require an exponential number of switches and so multiple stages appear to be useful. Also, although for simplicity we have assumed that the outputs of stage i are used directly only in stage i+1, they could equally be used in stages i+2, i+3, etc. We will not consider these additional complexities here because they are secondary to the fundamental issues.

LATCHES

What has been described so far will allow the construction of an N level reversible combinational logic circuit. To be useful, we must also specify a latch which can store the output of the circuit. This can be done by holding charge on a capacitor. The outputs of stage N are connected through switches to a latch stage, stage N+1.

The latch requires its own clock. Such a latch is shown in figure. Initially, the latch clock is 0 and allows charge to move freely into the capacitor. When the output from stage N is valid, the latch clock goes high (changes to logic 1) and prevents charge from escaping.

Clock N can now go low (to logic 0) which will make the output from stage N also go low, but the output of the latch will not be effected by changes in the output from stage N because it is cut off. Of course, having once "locked up" the outputs in the latches, we cannot release the information in the latches. If we should again set the clock latch low, allowing charge to flow off the capacitor, then we would be turning on a switch when there was a voltage across it. This is forbidden, and so the problem of removing information from a latch is non-trivial.

UNWRITING

Rather than simply "erasing" the information on the latch by letting the latch clock go low, we need to "unwrite" the information. Unwriting differs fundamentally from erasing. When we erase, we do not know the state of the latch, we simply turn on the switch holding the charge on the latch and let the charge flow away.

Whether the latch held a "0" or a "1" is immaterial. In the process, we must necessarily dissipate energy because erasing information is irreversible. By contrast, when we "unwrite" a latch we actually have an input signal that tells us the state of the latch. We can set the latch clock low and connect the input signal and the value stored on the latch.

Because (by definition) the input signal is in the same logical state as the latch, there is no voltage across the switch when we turn it on. We can now let the input signal go to 0 slowly, thus draining charge from the latch.We can indefinitely reduce the energy dissipated by this process by slowing the speed with which we drain charge from the latch.

Summary of Combinational Logic

In our earlier description, we left the source of valid input a mystery. The outputs of stage 0 held valid data, and how this happy state of affairs came about was unspecified. Now we can rectify this omission: stage 0 is a latch. This implies that clock 0 must be kept high to prevent the charges in stage 0 from leaking away.

Further, stage N+1 is a latch, and holds the output. We still leave the mechanism that initially put proper charge levels into the capacitors of stage 0 a mystery, but it is now clear that stage 0 will continue to hold a valid signal until clock 0 is lowered and the charge is allowed to escape.

The latches of stage 0 will also be called the input latches or simply the input, while the latches of stage N+1 will also be called the output latches or simply the output.

We can now specify our initial state as one in which appropriate charges are held on the input, clock 0 is high (to prevent charge from the input latches from leaking away), while clocks 1 through N are held low; as is clock signal N+1 (that is, there are no charges held on the output latches and the switch connecting the output latches to the final logic stage is turned on).

We raise clock signals 1 through N in turn, computing the values of each logic level in turn. Finally, we raise clock signal N+1, thus trapping the output signals in the output latches. We are then free to lower clock signal N, then N–1, then N–2, etc. until we have finally worked our way back to the first logic level. When the computation is done, all clock signals from 1 through N are held low, while clock signals 0 and N+1 remain high. Charge is trapped in stage 0 corresponding to the input, and charge is trapped in stage N+1 corresponding to the output. The outputs of stage 0 and N+1 are valid. All other outputs are 0.

ITERATION OF A REVERSIBLE LOGIC FUNCTION

At a higher level of abstraction, we can say that the outputs of stage N+1 are equal to some combinational function F applied to the outputs of stage 0, or output = F(input). Of course, we're now stuck. We cannot unwrite the input latches because we don't know what information is stored in them. We can solve this dilemma by requiring that F be reversible. That is, we demand that F^–1(F(input)) = input for all possible values of the input on stage 0. This allows us to compute the values stored in stage 0 from the values stored in stage N+1. We need merely compute F^–1(output).

That is, we add additional logic stages N+2, N+3..., N+M which compute the inverse of F. The output of stage N+M will then be the same as the input. Because we now have data values which are identical to the contents of stage 0, we can unwrite stage 0 in a fully reversible fashion.

That is, we can lower clock 0 and allow the charge on the capacitors in stage 0 to be connected with the output signals from stage N+M. Because they are identical the energy loss from this operation can be made arbitrarily small.

Then, clock signal N+M can be lowered, clock signal N+M–1 can be lowered, etc. until we finally lower clock signal N+2. Clock signal N+1 (the latch control clock signal for the output) is high (to hold the output steady in the output latches) and remains high.

To review the cycle of operations from the initial state: we start with a valid input in stage 0, with clock signal 0 held high (to hold the charge in the latches of stage 0) and with all other clock signals held low. From this intial state, we start the computation by raising clock signals 1 through N in turn, computing F(input). Clock signal N+1 is then raised to hold the output values in the output latches of stage N+1, and then clock signals N, N–1, N–2..., 1 are lowered in their turn to "uncompute" the calculation of F. We then compute F ^ -1(output) by raising clock signals N+2, N+3, N+4..., N+M. This computation produces exactly the values that are stored in stage 0, and so we can now unwrite the contents of stage 0 by lowering clock 0: this turns on the switches connecting the latches of stage 0 to the output of stage N+M.

We then uncompute the calculation of F^{-1}(output) by succesively lowering clock signals N+M, N+M–1, N+M–2..., and N+2. Finally, we again raise clock 0 to again isolate the input (which is now all 0). At the end of this process the outputs are stored in stage N+1 and the input is all 0. The entire process was carried out in a fully reversible manner, and so energy dissipation can in principle be made arbitrarily small.

The only thing we need finally do is exchange the data in the input and output latches (we assume that the input and output latches have the same number of bits). To do this, we first transfer the data in the output latches into the input latches and then "unwrite" the output latches. While the circuitry for this is not entirely trivial, it should be clear that the principles developed should allow us to do this. The sequence of steps involved is gone over in detail in the following several paragraphs.

To exchange the data between the input latches and the output latches, we require four additional sets of switches and four additional sets of clock lines. We illustrate the process by considering the exchange between a single input latch and a single output latch. Initially, EC_1 and EC_2 are set to 0 while EC_3 and EC_4 are set to 1. The switches S_3 and S_4 are turned off. S_1 is turned on, because the input latch holds a 0. S_2 could be either on or off, depending on the data on the output latch.

We first transfer data from the output latch to the input latch. EC_3 is lowered turning on S_3. Because the input latch holds 0 and EC_2 is set to 0, turning on S_3 is permitted. EC_2 is then raised to 1. If S_2 is on, charge is transferred to the input latch. If S_2 is off, charge is not transferred.

Thus, if the output latch holds a 1 the input latch will be set to 0. If the output latch holds a 0, the input latch will be set to 1. In the process of transferring the data from the output latch to the input latch, we have also inverted it. To correct this omission would require an inverter in the diagram of figure, or some modifications to F and F^{-1}. This would take an additional switch and an additional clock line, as well as additional explanation. The figure, however, would rapidly become cluttered so we have chosen to omit the inverter to clarify the exposition. This ommision has no impact on the basic concept.

Having set the input latch with data from the output latch, we now raise EC_3 and turn off S_3. Finally, we lower EC_2. This completes the transfer of data from the output latches to the input latches. We must now unwrite the output latches. Unwriting the output latches is similar to setting the input latches, but the exact sequence of operations is changed. Our first step is to raise EC_1. We then lower EC_4 turning on S_4. EC_1 is high, so if there is charge on the output latch then we have turned on a switch when there is no voltage across it, which is permitted.

If there is no charge on the output latch, then there will be charge on the input latch (because we transferred the complement of the output latch to the input latch in the earlier steps) and this charge will turn off S_1. Therefore, the output latch and EC_1 will not be connected. Again, we have not turned on a switch with a voltage across it.

At this point, if the output latch is charged it is connected to EC_1. By lowering EC_1 we can drain the charge from the output latch. We then raise EC_4 turning off S_4. This isolates the output latch again and completes the unwriting of the output latch from the input latch.

The clocking sequence, in brief, is: lower EC_3, raise EC_2, raise EC_3, lower EC_2, raise EC_1, lower EC_4, lower EC_1, raise EC_4. This sequence of operations has exchanged the data on the input latch and the output latch, and did not require turning on a switch when there was a voltage across it. The careful reader might wonder about the charge that might be trapped on the segment between S_1 and S_4 or between S_2 and S_3. These segments are initially 0, and are restored to 0 upon completion of all operations.

During the actual transfer and unwriting steps, the charge on these segments is always known and does not result in turning on a switch with a voltage across it. An explicit capacitance attached to this segment would not alter the conclusions reached here for this specific design.

In other designs it might be useful to eliminate such segments. Whether or not such segments can be eliminated depends on the specific implementation. Elimination can often be done by appropriate physical design of the switches, effectively resulting in a single switch with two control inputs.

We have now completed the full cycle of operations required to compute F(input). We can go on to compute F(F(input)), F(F(F(input))), etc. We assume F is computationally useful (*e.g.*, it implements a single instruction or step of a reversible computation). As discussed earlier, reversible functions can be used to carry out arbitrary computations.

PROBLEMS WITH INACCURATE VOLTAGE

While the system described is reversible in principle, the energy dissipation in a practical implementation would depend upon the accuracy of the voltage. When a switch is turned on, the logical state of the input and output values is the same, but the physical state might differ. A small difference in the voltages at the input and output of the switch would result in energy loss as the switch was turned on. Even with this restriction, it should be possible to achieve significant reductions in energy dissipation. The energy dissipated by dissipatively discharging a capacitive load is,

$$\text{Energy} = 1/2\ CV^2$$

where C is the capacitance and V is the voltage. In conventional (irreversible) logic systems, the voltage V would be the voltage of the power supply. In a reversible system of the type described here the voltage V would be the error in the voltage of the power supply, rather than the voltage of the power supply itself. Thus, by using reversible instead of irreversible logic, we have effectively reduced the value of V. With a power supply accurate to 5 per cent, the energy losses from charging and discharging the load capacitances associated with the electrical connections would be decreased by a factor of 20^2 or 400. That is, energy losses would be a function of the error in the voltage of the power supply, rather than a direct function of the voltage of the power supply itself.

There does not appear to be any reason in principle why the error in the voltage of the power supply could not be made as small as desired. A more fundamental reason for long term optimism is the observation that electric charge is quantized. In future systems in which the number of electrons that represents a logic "0" or "1" is both finite and small, it should be possible to reduce the energy loss caused by slight inaccuracies in the voltage supply to 0.

It is already possible to control the passage of individual electrons in a circuit. Such highly precise control should eventually make a fully reversible electronic switch with extremely small energy losses feasible.

Problems with FETS: Voltage Degradation

Switching elements that do not degrade the voltage of the signal being switched are feasible in principle, as shown by the electrostatic relay illustrated in figure. CMOS transmission gates can also be used as switches without concern about voltage degradation. Transmission gates require that both the control signal and its inverse be provided. A simple method of satisfying this requirement would be to use two-rail logic. To use nMOS (or pMOS) by itself, we must overcome certain shortcomings.

In particular an nMOS switch, when used in pass logic, will cause voltage degradation. As a consequence, deterioration of the voltage through multiple logic levels would impose a severe limit on the capabilities of the system. Boosting voltage in a fully reversible fashion is, however, straightforwards in principle. The voltage on a capacitor will increase if the capacitance decreases. Variable capacitors are well known based both on physical motion of the capacitor plates, or on changes in charge distribution secondary to a voltage change. Varactors in which the capacitance C changes by a factor of 10 or more in response to a voltage change of little more than a volt are feasible. The approach outlined here is abstractly similar to the proposal by Joynson et. al.

Boosting voltage is illustrated in figure. Before starting, the switch is turned on and current can flow onto the varactor. The high-voltage clock then opens the switch, trapping the charge on the varactor. Finally, the capacitance of the varactor is reduced. Because the voltage on a capacitor is $V=Q/C$ (where V is voltage, Q is charge and C is capacitance) the voltage across the varactor will increase as its capacitance decreases. Because the voltage on the high-voltage clock is large, there is more than sufficient control voltage to insure that the switch remains turned off even though the voltage across the switch has been increased. The low-voltage input, boosted to a medium-voltage output, can then be used to switch a low-voltage signal, which can again be used as the input to a reversible voltage booster. To reverse this cycle of operations, the capacitance of the varactor is returned to its original value.

This will also return the voltage on the output to its original value, and the high-voltage switch can be turned on with minimal energy loss. This energy loss will be a function of the square of the error in the voltage regulation. If the number of electrons that charge the varactor (and any stray output capacitance) is reduced to a small integer number, it should eventually be possible to eliminate losses due to small inaccuracies in voltage regulation. Seitz et. al. commented that "...the isolation transistor can turn on while there is voltage across it, and accordingly, it dissipates power in charging or discharging the bootstrap node. The goal of exporting all of the dynamic power is elusive." Translating from their language to the language used here, they sometimes allowed the high-voltage clock to turn on the high-voltage switch when a voltage existed across the switch, thus dissipating energy.

They did not consider the computation of F ^ -1 and "unwriting" operations, and so were sometimes obliged to engage in the dissipative operation of discharging the gate of a FET to ground. In essence, they were able to eliminate unnecessarily dissipative operations for combinational logic, but not for sequential logic. As shown here, appropriate design of a reversible sequential circuit eliminates the unnecessarily dissipative steps.

PROBLEMS WITH FETS

Leakage Current

Another problem with FET's is leakage current. Even a FET in which the gate voltage is sufficient to insure that the FET is completely cut off will still pass a small current. If our objective is to achieve energy dissipations below kT, any significant leakage current is intolerable. Leakage current occurs because thermally generated electron-hole pairs are created in the semiconductor material of the FET.

Thermal agitation will occasionally cause an electron in the valence band (*e.g.*, an electron which normally would not be able to move because it was fixed in place as part of a covalent bond) to leap into the conduction band. An electron in the conduction band is free to move throughout the crystal and so to conduct electric current.

The energy gap between the valence band and the conduction band will govern the concentration of electrons that are available to carry leakage current. In silicon, the bandgap is 1.12 electron volts.

At 300 kelvins, this will produce 1.45×10^{10} electrons per cubic centimeter of silicon. This "intrinsic carrier concentration" will result in dissipative energy losses.

Three obvious methods are available for reducing the leakage current:

- *Reduce the physical size of the device*: An intrinsic carrier concentration of 1.45×10^{10} electrons per cubic centimeter can also be expressed as 1.45×10^{-11} electrons per cubic nanometer. If the active region of the device was smaller than a few millions of cubic nanometers, then the probability of finding an undesired charge carrier in that region would be relatively small.
- *Reduce the temperature*: *The intrinsic carrier concentration is*:

  $$ni = km \exp(-Eg/2kt)$$

  where ni is the intrinsic carrier concentration, km is a constant specific to the material, and Eg is the bandgap energy. In the case of silicon, km is about 3.1×10^{20}. At a temperature of 77 Kelvins, ni is 7.4×10^{-17} electrons per cubic centimeter. CCD's with dark currents of a few electrons per pixel per hour have been fabricated.
- Select a material with a larger bandgap. Diamond has a bandgap of 5.47 electron volts, with an intrinsic carrier concentration at room temperature in the range of 10^{-26} electrons per cubic centimeter. Boron provides a good shallow dopant, and p-type diamond has been made. Several shallow dopants for n-type material are plausible. Diamond MOSFETs with boron doping have been fabricated. An implementation could use purely p-type materials, if that proved useful.

  A variety of wide band-gap materials exist allowing a wide range of tradeoffs between energy dissipation and other parameters. It should be feasible to produce acceptable leakage currents using a variety of different materials in a variety of different geometries under a broad range of operating temperatures. The use of small FET's using a wide band-gap material is particularly attractive for room temperature operation. In such a structure the probability of finding even one valence electron in the conduction band is astronomically small.

Thermally generated leakage current is also of concern in many types of varactors. A reverse- biased P-N junction illustrates the problem: electron-hole pairs generated thermally in the charge depletion region are swept out of the region, thus conducting current. Leakage current in such varactors can in principle be made quite small for the same reasons that apply to leakage current in a FET.

Reversible CCD-Based Logic

An alternative approach, which eliminates the need for high accuracy voltage supplies, is to eliminate the need to turn on the switch when there are charge carriers present. This might be viewed as only turning the switch on and off when the voltage across the switch is 0 volts, but this is not a fully adequate description. Even a 0 volt signal is subject to fluctuation and noise. If, however, we try to put a positive voltage onto the switch (*e.g.*, the inputs and outputs of the switch both have a positive voltage), and if the charge carriers in the switch are electrons, then it is possible to remove all the charge carriers from the channel region.

Any charge carriers will be actively swept out of the channel region by the positive voltages on the input and output leads. Once all the charge carriers have been removed from the channel region, then no more current will flow. Precise regulation of the positive voltage on the input and output of the switch is not required, for both need merely be "sufficiently positive" to insure removal of all charge carriers from the channel region. Once the charge carriers have been removed then turning the switch on and off can be done in a fully reversible manner. Small errors in the supply voltage can no longer lead to dissipative current flows when the switch is turned on, because no current can flow.

For purposes of this discussion we will assume our switch is a FET with electrons as the charge carriers and that the electrons are removed from the FET by a positive voltage on the input and output leads. (We could equally well have assumed the charge carriers are holes and that a negative voltage is placed on the input and output leads: for purposes of explanation the choice is arbitrary). As discussed earlier, it is essential that the device be designed so that the concentration of intrinsic charge carriers is sufficiently low that energy losses from this source can be neglected. Further, the density of device defects must be low enough that defect sites do not appreciably alter device performance.

We will simply assume that the material is free of defects. Reduction of device size coupled with decreases in defect rates should eventually allow construction of devices with a sufficiently low defect rate that energy dissipation will not be significantly degraded. In this approach, 1's and 0's are represented by a packet of electrons or the absence of a packet of electrons, respectively. The electrons are stored in potential wells which we will call "buckets." Each packet of electrons is treated as a unit: we never split or merge packets, nor do we create or destroy packets (except when the system is initialized, which would ideally occur only once). While all packets have approximately the same number of electrons, it is not critical that each packet have exactly the right number of electrons. Whether a packet has 99, 100, or 101 electrons won't matter.

The use of charge packets is most familiar in the context of CCDs (Charge Coupled Devices), or alternatively in the context of BBD's, or Bucket Brigade Devices. In a CCD, charge packets are transferred from potential well to potential well serially along a row of devices. The amount of charge in each packet is a measure of some analog quantity. Such devices are primarily used as memory elements or shift registers, not for computation. Proposals to use the same basic concepts for logic operations have been made since the invention of the CCD.

Sometimes called "DCCL," or Digital Charge Coupled Logic, these proposals have explored the advantages of using charge packets as the basic logic element. The primitive logic elements in prior proposals, however, were highly dissipative. For example, Zimmerman et. al. proposed logic operations whose computation required merging packets, discarding packets by draining the charge to ground, etc.

The goal of thermodynamically reversible logic was not an objective. Tompsett proposed circuits in which the potential generated by a charge packet in a CCD potential well was coupled to the gate of a FET. This FET would then be turned on or off depending on whether a charge packet was present or absent in the potential well. This was then used to create another charge packet (rather than to control the movement of an existing charge packet). The creation of a new charge packet would compensate for imperfect charge transfer.

NAND and NOR operations could be performed by applying the potential from two potential wells to two gates of two FETs, which could then be connected in serial or parallel. Again, the objective of thermodynamic reversibility was not considered. The method of sensing charge in the potential well was dissipative and the sensed packets, no longer needed, were to be dissipatively discarded. For simplicity, we will assume that packets never gain or lose electrons (due to defects in the material or to thermal noise).

In fact, present day devices can gain or lose charge at the rate of a few parts per million. As discussed earlier, such gains or losses should eventually be reduced to insignificant levels. Even if we allow packets to gain or lose electrons at some low rate, we could periodically "refresh" the packets, and so prevent malfunction. Tompsett first proposed methods of refreshing packets (though he did not consider reversible methods). To the extent that the packet has gained or lost an unpredictable amount of charge, "refreshing" the packet must fundamentally dissipate energy. Other than this fundamental loss, however, we wish to keep energy dissipation to a minimum.

A reversible method of refreshing a packet would be to use the old packet to charge a bucket in a reversible fashion (as discussed in the first method), and then to "unwrite" or discharge the bucket holding the old packet by using the new packet as the source of information. This method of refreshing a packet would, of course, dissipate energy as a function of the errors in the voltages involved (as discussed earlier). If the refresh operations were done infrequently (*e.g.*, a packet would be refreshed only after a large number of logic operations) then the energy dissipation per logic operation from this source would be quite small. While CCD and DCCL circuits are today implemented on a semiconductor surface using planar technology, for purposes of this discussion we will initially ignore this geometrical constraint and instead focus on more fundamental issues. A planar version of the ideas described here is given in the section on "A Planar Version of RCT Logic."

Illustrates the "hydraulic" model of charge transfer in a CCD. Charge is transferred from one potential well to the next much like water would flow to a gradually sinking region in a pond. This process can be thermodynamically reversible when appropriately implemented. One aspect of figure might be slightly misleading. While water has a sharply defined surface, the energy of an electron at some non-zero temperature T is statistical. While figure might lead one to conclude that extremely small perturbations in the potential might cause significant changes in the pattern of charge flow, in fact thermal noise will allow electrons to surmount small barriers *e.g.*, where the barriers have a potential similar in magnitude to kT/e or.026 volts at room temperature (where e is the charge on an electron). Small perturbations and errors in the potential do not fundamentally create dissipative problems.

TOXICOLOGY OF NANOPARTICLES

Nanotoxicology is a sub-specialty of particle toxicology. It addresses the toxicology of nanoparticles which appear to have toxicity effects that are unusual and not seen with larger particles. Nanoparticles can be divided into combustion-derived nanoparticles, manufactured nanoparticles like carbon nanotubes and naturally occurring nanoparticles from volcanic eruptions, atmospheric chemistry etc. Typical nanoparticles that have been studied are titanium dioxide, alumina, zinc oxide, carbon black, and carbon nanotubes, and "nano-C_{60}". Nanoparticles have much larger surface area to unit mass ratios which in some cases may lead to greater pro-inflammatory effects. In addition, some nanoparticles seem to be able to translocate from their site of deposition to distant sites such as the blood and the brain. This has resulted in a sea-change in how particle toxicology is viewed-instead of being confined to the lungs, nanoparticle toxicologists study the brain, blood, liver, skin and gut. Nanotoxicology has revolutionised particle toxicology and rejuvenated it.

Reactive Oxygen Species

For some types of particles, the smaller they are, the greater their surface area to volume ratio and the higher their chemical reactivity and biological activity. The greater chemical reactivity of nanomaterials can result in increased production of *reactive oxygen species* (ROS), including free radicals. ROS production has been found in a diverse range of nanomaterials including carbon fullerenes, carbon nanotubes and nanoparticle metal oxides. ROS and free radical production is one of the primary mechanisms of nanoparticle toxicity; it may result in oxidative stress, inflammation, and consequent damage to proteins, membranes and DNA.

Biodistribution

The extremely small size of nanomaterials also means that they much more readily gain entry into the human body than larger sized particles. How these nanoparticles behave inside the body is still a major question that needs to be resolved. The behaviour of nanoparticles is a function of their size, shape and surface reactivity with the surrounding tissue. In principle, a large number of particles could overload the body's phagocytes, cells that ingest and destroy foreign matter, thereby triggering stress reactions that lead to inflammation and weaken the body's defence against other pathogens.

In addition to questions about what happens if non-degradable or slowly degradable nanoparticles accumulate in bodily organs, another concern is their potential interaction or interference with biological processes inside the body. Because of their large surface area, nanoparticles will, on exposure to tissue and fluids, immediately adsorb onto their surface some of the macromolecules they encounter. This may, for instance, affect the regulatory mechanisms of enzymes and other proteins. Nanomaterials are able to cross biological membranes and access cells, tissues and organs that larger-sized particles normally cannot. Nanomaterials can gain access to the blood stream via inhalation or ingestion.

At least some nanomaterials can penetrate the skin; even larger microparticles may penetrate skin when it is flexed. Broken skin is an ineffective particle barrier, suggesting that acne, eczema, shaving wounds or severe sunburn may accelerate skin uptake of nanomaterials. Then, once in the blood stream, nanomaterials can be transported around the body and be taken up by organs and tissues, including the brain, heart, liver, kidneys, spleen, bone marrow and nervous system. Nanomaterials have proved toxic to human tissue and cell cultures, resulting in increased oxidative stress, inflammatory cytokine production and cell death. Unlike larger particles, nanomaterials may be taken up by cell mitochondria and the cell nucleus. Studies demonstrate the potential for nanomaterials to cause DNA mutation and induce major structural damage to mitochondria, even resulting in cell death.

Nanotoxicity Studies

Since there is no authority to regulate nanotech-based products, there are many products that could possibly be dangerous to humans. Scientific research has indicated the potential for some nanomaterials to be toxic to humans or the environment. In March 2004 tests conducted by environmental toxicologist Eva Oberdörster, Ph.D. working with Southern Methodist University in Texas, found extensive brain damage to fish exposed to fullerenes for a period of just 48 hours at a relatively moderate dose of 0.5 parts per million. The fish also exhibited changed gene markers in their livers, indicating their entire physiology was affected.

In a concurrent test, the fullerenes killed water fleas, an important link in the marine food chain. The extremely small size of fabricated nanomaterials also means that they are much more readily taken up by living tissue than presently known toxins. Nanoparticles can be inhaled, swallowed, absorbed through skin and deliberately or accidentally injected during medical procedures. They might be accidentally or inadvertently released from materials implanted into living tissue.

Researcher Shosaku Kashiwada of the National Institute for Environ-mental Studies in Tsukuba, Japan, in a more recent study, intended to further investigate the effects of nanoparticles on soft-bodied organisms. His study allowed him to explore the distribution of water-suspended fluorescent nanoparticles throughout the eggs and adult bodies of a species of fish, known as the see-through medaka. Moreover, small fish like the see-through medaka have been popular test subjects for human diseases and organogenesis for other reasons as well, including their transparent embryos, rapid embryo development, and the functional equivalence of their organs and tissue material to that of mammals.

Because the see-through medaka have transparent bodies, analysing the deposition of fluorescent nanoparticles throughout the body is quite simple. For his study, Dr. Kashiwada evaluated four aspects of nanoparticle accumulation. These included the overall accumulation and the size-dependent accumulation of nanoparticles by medaka eggs, the effects of salinity on the aggregation of nanoparticles in solution and on their accumulation by medaka eggs, and the distribution of nanoparticles in the blood and organs of adult medaka. It was also noted that nanoparticles were in fact taken up into the bloodstream and deposited throughout the body. In the medaka eggs, there was a high accumulation of nanoparticles in the yolk; most often bioavailibility was dependent on specific sizes of the particles. Adult samples of medaka had accumulated nanoparticles in the gills, intestine, brain, testis, liver, and bloodstream. One major result from this study was the fact that salinity may have a large influence on the bioavailibility and toxicity of nanoparticles to penetrate membranes and eventually kill the specimen. As the use of nanomaterials increases worldwide, concerns for worker and user safety are mounting. To address such concerns, the Swedish Karolinska Institute conducted a study in which various nanoparticles were introduced to human lung epithelial cells. The results, released in 2008, showed that iron oxide nanoparticles caused little DNA damage and were non-toxic. Zinc oxide nanoparticles were slightly worse. Titanium dioxide caused only DNA damage. Carbon nanotubes caused DNA damage at low levels. Copper oxide was found to be the worst offender, and was the only nanomaterial identified by the researchers as a clear health risk.

No Fullerene Toxicity Reported

Nanoparticles can also be made of C_{60}, as is the case with almost any room temperature solid, and several groups have done this and studied toxicity of such particles. The results in the work of Oberdörster at Southern Methodist University, published in "Environmental Health Perspectives" in July 2004, in which questions were raised of potential cytotoxicity, has now been shown by several sources to be likely caused by the tetrahydrofuran used in preparing the 30 nm–100 nm particles of C_{60} used in the research. Isakovic, *et al.*, 2006, who review this phenomenon, gives results showing that removal of THF from the C_{60} particles resulted in a loss of toxicity. Sayes, *et al.*, 2007, also show that particles prepared as in Oberdorster caused no detectable inflammatory response when instilled intratracheally in rats after observation for 3 months, suggesting that even the particles prepared by Oberdorster do not exhibit markers of toxicity in mammalian models.

This work used as a benchmark quartz particles, which did give an inflammatory response. A comprehensive and recent review of work on fullerene toxicity is available in "Toxicity Studies of Fullerenes and Derivatives," a chapter from the book "Bio-applications of Nanoparticles". In this work, the authors review the work on fullerene toxicity beginning in the early 1990s to present, and conclude that the evidence gathered since the discovery of fullerenes overwhelmingly points to C_{60} being non-toxic. As is the case for toxicity profile with any chemical modification of a structural moiety, the authors suggest that individual molecules be assessed individually.

Immunogenicity of Nanoparticles

Very little attention has been directed towards the potential immunogenicity of nanostructures. Nanostructures can activate the immune system inducing inflammation, immune responses, allergy, or even affect to the immune cells in a deleterious or beneficial way. More studies are needed in order to know the potential deleterious or beneficial effects of nano-structures in the immune system. In comparison to conventional pharmeceutical agents, nanostructures have very large sizes and immune cells, especially phagocytic cells, recognise and try to destroy them.

Complications with Nanotoxicity Studies

Size is therefore a key factor in determining the potential toxicity of a particle. However it is not the only important factor. Other properties of nanomaterials that influence toxicity include: chemical composition, shape, surface structure, surface charge, aggregation and solubility, and the presence or absence of functional groups of other chemicals. The large number of variables influencing toxicity means that it is difficult to generalise about health risks associated with exposure to nanomaterials–each new nanomaterial must be assessed individually and all material properties must be taken into account.

In addition, standarization of toxicology tests between laboratories are needed. Díaz, B. *et al.* from the University of Vigo has shown that many different cell lines should be studied in order to know if a nanostructure induces toxicity, and human cells can internalise aggregated nanoparticles. Moreover, it is important to take into account that many nanostructures aggregate in biological fluids, but groups manufacturing nanostructures do not care much about this matter. Many efforts of interdisciplinary groups are strongly needed in order to progress in this field.

Effect of Aggregation/Agglomeration of Nanoparticles

Many nanoparticles agglomerate or aggregate when they are placed in environmental or biological fluids. The terms agglomeration and aggregation have distinct definitions according to the standards organizations ISO and ASTM, where agglomeration signifies more loosely bound particles and aggregation signifies very tightly bound or fused particles. Nanoparticles frequently agglomerate due to the high ionic strength of environmental and biological fluids, which shields the repulsion due to charges on the nanoparticles. Unfortunately, agglomeration has frequently been ignored in nanotoxicity studies, even though agglomeration would be expected to affect nanotoxicity since it changes the size, surface area, and sedimentation properties of the nanoparticles.

In addition, many nanoparticles will agglomerate to some extent in the environment or in the body before they reach their target, so it is desirable to study how toxicity is affected by agglomeration. A method was published that can be used to produce different mean sizes of stable agglomerates of several metal, metal oxide, and polymer nanoparticles in cell culture media for cell toxicity studies. Different mean sizes of agglomerates are produced by allowing the nanoparticles to agglomerate to a particular size in cell culture media without protein, and then adding protein to coat the agglomerates and "freeze" them at that size.

By waiting different amounts of time before adding protein, different mean sizes of agglomerates of a single type of nanoparticle can be produced in an otherwise identical solution, allowing one to study how agglomerate size affects toxicity. In addition, it was found that vortexing while adding a high concentration of nanoparticles to the cell culture media produces much less agglomerated nanoparticles than if the dispersed solution is only mixed after adding the nanoparticles.

Challenges of the Nano-Visualisation and Related Unknowns in Nanotoxicology

With comparison to more conventional toxicology studies, the nanotoxicology field is however suffering form a lack of easy characterisation of the potential contaminants, the "nano" scale been still a scale difficult to apprehend. The biological systems are themselves still not completely known at this scale. Ultimate Atomic visualisation methods such as Electron microscopy and *Atomic force Microscopy* (AFM) analysis are allowing fantastic progresses in the visualisation of the nano world.

Yet, further nanotoxicology studies will require extremely precise characterisation of the specificities of a given nano-element: size, chemical composition, detailed shape, level of aggregation, combination with other vectors, etc. These properties would have to be determined not only on the nanocomponent before its introduction in the living environnment but also in the biological environnement. This is why nanotoxicoly is a fantastic field of research. This is also why it is not easy to determine to what extent a given nanoparticule has a dramatic effect when compared to comparable nanoparticules already present in our environnement either through natural/ biological origin or through ancestral human activity.

TRANFERRING CHARGE WITH LITTLE DISSIPATION

Before discussing the more complex operations needed for computation, we first consider a simple sequence that transfers charge from a source bucket to a destination bucket with as little energy loss as desired. This is illustrated in figure. (Note that the illustration shows that the two buckets have semiconductor plates joined by a semiconductor path. The use of metallic plates or a metallic connection between the two plates would cause unwanted energy dissipation). At first, both the source clock and destination clock are + (positive), and the charge packet is entirely in the source bucket. The clocking sequence will then shift the charge from the source bucket to the destination bucket in a series of steps, where each step can dissipate an arbitrarily small amount of energy. As we move from state 1 to state 2, the destination clock is changed from + to -. This change results in negligible energy dissipation, because no charge can move at this time. As we move from state 2 to state 3, the source clock is changed from + to -. The charge carriers in the charge packet (electrons in our example) now move out of the source bucket. Because the destination clock is negative, the electrons will migrate smoothly out of the source bucket. (If the destination clock were + at this point instead of -, the electrons would move out of the source bucket and then "fall down hill" into the destination bucket, dissipating energy). As we move from state 3 to state 4, the destination clock changes from-to +, thus slowly attracting the electrons and gathering them into the destination bucket. Finally, as we move from state 4 to state 5, the source clock changes from-to + restoring the system to its original condition, but with the charge packet in the destination bucket instead of the source bucket.

The primary source of energy dissipation in this sequence is from state 2 to state 3, and from state 3 to state 4. These transitions involve charge migration, and will therefore result in dissipative losses. As we slow the clock frequency the clock voltages change more slowly, the charges move more slowly, and hence dissipate less energy. Because energy dissipation is a function of the square of the current, reducing the frequency of operation by a factor of two will also reduce the current by a factor of two but will reduce the energy dissipation by a factor of 4. This will reduce the energy dissipated per logic operation by a factor of 2. Thus, by sufficiently slowing the clock frequency we can dissipate as little energy per operation as might be desired. It is important to notice that these operations do not erase information. We start with a charge packet in a single potential well on the left, and we end with a charge packet in a single potential well on the right.

During the course of moving the charge packet we slowly change the shape and position of the potential well, but at no point do we merge potential wells. This transition eliminates the right hand potential well without merging it with the well on the left. An attempt to merge the two potential wells would be dissipative. The transition from step 2 to step 3 slowly changes the shape of the potential well, but the charge packet is always contained in a single potential well and is always at or near thermodynamic equilibrium within that well.

THE PRIMITIVE OPERATION

We now consider the somewhat more complex operation that is required to support computation in a reversible fashion.

The single primitive operation is: Transfer charge out of a source bucket and into an empty destination bucket if a "condition" bucket holds a "0", but do not transfer the charge if the condition bucket holds a "1". If the transfer does not take place (the condition bucket is "1"), the destination bucket need not be empty. "Empty" in this context means "has no charge carriers present." That is, an "empty" bucket is in logic state "0." We will call this device a 3-bucket Reversible Charge Transfer device, or (where context makes it clear which type of RCT we are referring to) simply an RCT.

The RCT is a reversible DCCL device. At a logical level the RCT is related to (but somewhat different from) a switch gate implemented using charge packets instead of colliding billiard balls. The particular logic function described here is but one example of the range of logic functions that involve the reversible transfer of charge from one bucket to another.

During normal operation, RCT devices:

- Never merge or split charge packets
- Never discharge charge packets to ground, and
- Move charge packets in a thermodynamically reversible fashion from a set of source buckets to a set of destination buckets, conditional on the presence or absence of charge in a set of condition buckets.

In this context thermodynamic reversibility means that energy dissipation per device operation, in the absence of manufacturing defects, could be made much smaller than kT simply by slowing the speed of operation.

The number of charge packets is conserved and therefore RCT devices are conservative logic devices (in conservative logic devices, the number of 1's is conserved during the computation, so the number of 1's at the input is the same as the number of 1's at the output).

We have described the 3-bucket RCT, which has one condition bucket, one source bucket, and one destination bucket. More complex RCT's have more buckets and the pattern of charge transfer is more complex. Some of these RCT's will be considered later.

The simple 3-bucket RCT operation might be written in C as:

- If (Condition==0) {Destination = Source; Source = 0}; while a somewhat briefer notation would be:
- If Condition is 0 then Destination = Source. (Note that in this briefer notation, the "=" sign is used to indicate both the assignment of the Source value to the Destination, and setting the Source to 0 after the assignment).

The precondition that must be true prior to execution of the 3-bucket RCT is:

(Destination==0) OR (Condition==1)

Violation of this precondition would result in an unwanted dissipative step, which is banned in RCT devices. While it might at first seem that satisyfing this precondition would make the 3- bucket RCT data dependent, it is in fact possible to design a circuit in which we know that the precondition is satisfied without knowing what values are presented as inputs to the circuit. A logic diagram showing the possible initial and final states of a 3-bucket RCT is shown in Figure. On the left, the legal initial states of the RCT are shown, while on the right the final states (after the sequence of clock signals that cause the device to step through a single cycle of operations) are shown.

It is interesting to note that there is only a single legal, non-trivial change of state: when the input is 1, the condition is 0, and the output is 0. In all other cases either the input and output states are the same or the input state is illegal. The illegal states would cause undesired energy dissipation.

FREDKIN GATE WITH 3-BUCKET RCT

The 3-bucket RCT is logically complete, for we can implement a Fredkin gate with it. The construction used here is related to (though diffrent from) that used by Fredkin and Toffoli for constructing a billiard ball Fredkin gate from a billiard ball switch gate.

A Fredkin gate has three inputs and three outputs. The inputs are the control input A, and two signal inputs B and C. The outputs are A', B', and C'. If the control input A is 0, then A' = A, B' = B, and C' = C. If the control input A is 1, then A' = A, B' = C, and C' = B. That is, the output of the Fredkin gate is identical to the input if A = 0, but B' and C' are exchanged if A = 1.

The following sequence of simple RCT operations will implement a Fredkin gate: A, B, and C hold arbitrary input values. InitiallyOne is set to a logic "1" (the presence of an electron packet). All other variables are initialized to logical "0" (the absence of an electron packet).

- If A is 0 then C' = C.
- If A is 0 then B' = B.
- If A is 0 then NotA = InitiallyOne.
- If NotA is 0 then C' = B.
- If NotA is 0 then B' = C.
- If AlwaysZero is 0 then A' = A.
- If A' is 0 then InitiallyOne = NotA.

Steps 1 and 2 simply transfer charge from B and C into B' and C' if A is 0. B' and C' are initially empty, so this transfer satisfies the RCT preconditions.

Step 3 computes the logical negation of A and leaves that logical negation in "NotA."

Steps 4 and 5 transfer charge from B and C into C' and B', exchanging the outputs. The precondition is satisfied, for if A is 0, then NotA will be 1 and the transfer will not take place. If the transfer does not take place, the prior contents of B' and C' are irrelevant.

If A is 1, then NotA will be 0 and the transfer will take place. In this case, B' and C' will be empty (have logic values 0) and so the transfer will take place correctly.

Step 6 copies A into A' unconditionally.

Step 7 restores the value of InitiallyOne if it was altered during the computation. This implementation of a Fredkin gate will take three arbitrary input values and, after a sequence of RCT operations, produce three output values. We could clearly apply N Fredkin gates to 3N input values and produce 3N output values at the same time. We require that all output values be initially 0. After finishing the sequence of clocking operations, the inputs are all 0 and the outputs hold arbitrary values. At this point, the "input" and "output" can be logically exchanged (*e.g.*, no physical operation takes place, but we relabel the "input" variables as "output," and the "output" variables as "input"). Following this logical exchange, we have re-established the preconditions for the next sequence of operations: the (just relabeled) inputs again hold arbitrary logic values, while the (just relabeled) outputs hold logic 0's.

We now drive a (different) set of clock lines to apply the (different) Fredkin gates to generate the next "state" of the computation. Viewed globally, this is a "ping-pong" or "double buffer" scheme for computing successive values of the global state of the system. If we call the two sets of variables A and B, and we call the two logic functions for the "next state" of the system F and G, then we first compute B=F(A). B is initially all zeroes, while A holds arbitrary logic values.

On completion of the computation of F, A is all zeroes while B holds arbitrary logic values. We then compute A=G(B). Because G uses B as its input and A as its output, the condition that the outputs be all zeroes is met during the compuation of G. This returns us to the state where A holds arbitrary logic values and B holds all zeroes. We can repeat this cycle indefinitely.

IMPLEMENTATION OF A 3-BUCKET RCT

A planar layout of a 3-bucket RCT is shown in figure. Many kinds of implementations would be feasible, for example Silicon-On-Insulator technology (the FET used for the transfer switch would be "tilted on its side" in this approach). Note that many of the conductive paths are really semiconductors. Each bucket also has one semiconductor plate. The gate is also a semiconductor. Use of metallic wires for these components would compromise device function. A planar RCT is described in a later section.

This RCT can be viewed as a combination of three CCD potential wells and a FET. A charge packet is first transferred from the "condition" potential well to the gate of the FET. The charge packet in the "source" potential well is then moved to the source of the FET and (the charge on the gate of the FET permitting) through the channel and out the drain of the FET into the "destination" potential well.

Tompsett proposed sensing the presence or absence of a charge in a CCD potential well by connecting the potential well via a doped region connected ohmically to a metalization that became the gate electrode of a FET. Charge injection from the gate into the doped region, while small if the capacitance of the gate is small, still introduces an unwanted source of energy dissipation. Other aspects of Tompsett's proposal were also irreversible or highly dissipative. In the present proposal, we require that both the gate of the FET and the connecting path from the potential well to the gate be a semiconductor. The only charge carriers in this semiconductor path would be those deliberately introduced as a charge packet.

An alternative approach which might be more convenient in a simple planar implementation would be to introduce a capacitive coupling into the path from the potential well to the gate. The gate could then be metallic, one plate of the capacitor (the plate connected to the gate) could also be metallic, but the other plate and the connection to the potential well could be a semiconductor.

Potential could be coupled through the capacitor, but charge carriers would be prevented from moving into or out of the semiconductive region and hence charge injection from the gate electrode would not be a problem. Initially, a charge of unknown value is present in the source bucket. The destination bucket is either empty (if the condition bucket holds a 0) or has an unknown charge (if the condition bucket holds a 1). Clocks 1, 2, and 3 are all positive. All charge carriers (electrons) have been drawn onto the semiconductor plates of the buckets.

The conventional plates are connected to the clocks. There is a net positive charge on each bucket creating a potential well which holds the electrons in the bucket and prevents them from migrating away. The channel regions of all switches have no charge carriers present. Clock 4 is driven negative with a large voltage. This large voltage cuts off and isolates the three buckets involved from any other buckets to which they might normally be connected. Additional switches controlled by clock 4 might be required to further isolate the three buckets in a larger circuit, but are not shown in this small example.

The two gray channels do not actively participate in the process of reversible charge transfer, they are present merely to show that connections to other circuit components will be present in a real system. Clock 3 is driven negative, forcing charge from the condition bucket onto the gate of the transfer switch. If no charge is present on the condition bucket, no charge will be put onto the transfer gate, while if charge is present then it will be transferred to the transfer gate, cutting off the transfer switch. Note that the circuit is so arranged that only a medium voltage will be developed on the transfer gate, and this medium voltage can be contained by the high voltage of the cut off clock, clock 4.

Clock 2 is driven negative. If there is charge on the destination bucket, then the transfer switch is turned off and the charge on the destination bucket goes nowhere. If there is no charge on the destination bucket, then whether the transfer switch is turned off or on is immaterial, for no charge will be transferred.

Clock 1 is driven negative. If the transfer switch is turned off, no charge will be transferred. If the transfer switch is turned on, charge will move gradually through the transfer switch into the destination bucket. Because clock 2 is negative, charge will not "fall down hill" in a dissipative fashion when moving into the destination bucket.

Clock 1 has a low voltage so a low voltage is generated during the transfer operation. The medium voltage on the transfer gate is sufficient to block charge transfer if charge transfer is not supposed to take place.

Clock 2 is driven positive, allowing charge to flow smoothly into the destination bucket. When clock 2 reaches its full positive value, all the charge is held on the destination bucket and no charge carriers are present in either the source bucket or the transfer switch channel region (if the transfer switch was turned on). Clock 1 is driven positive, which has no effect if the transfer took place and simply allows the charge to return to its original state if the transfer did not take place.

Clock 3 is driven positive, allowing the charge on the transfer gate to return to the condition bucket. Clock 4 is driven positive, opening the cut off switches that isolated the three buckets from outside influence during the reversible transfer operation. This cycle of operations reversibly transfers charge from the source bucket to the destination bucket if the condition bucket holds a logic "0" (*e.g.*, no charge).

Energy Loss

There are losses during the RCT cycle of operations caused by current flowing through a resistive medium. The power lost will be, Power Loss = I^2 R where I is the current and R the resistance. Current is proportional to the frequency of operation of the circuit, so if the frequency is low the current will also be low. Therefore, power losses from resistive losses will fall off with the square of the frequency: Power Loss is proportional to f^2. Because the number of operations per second is also lower at a lower frequency, the actual energy loss per device operation is proportional to the frequency of operation: Energy Loss is proportional to f. By reducing the frequency of operation, the energy loss per operation can be reduced to whatever extent is desired. It should be remembered that "low frequency" operation is relative: CCD's with 0.9997 charge transfer efficiency operating at 1 gigahertz have been demonstrated.

More specifically, the electron drift velocity vd equals the electric field E times the mobility m:

$$vd = Em$$

If we let d be the distance traveled by an electron as it moves from one bucket to the next, and t be the time of a single logic operation, then it is approximately the case that:

$$d = vd\ t$$

The total energy dissipated by the movement of n electrons as they move from the source to the destination is simply the force times the distance times the number of electrons, or:

Edissipated-neEd

where Edissipated is the energy dissipated, n is the number of electrons in a packet, and e is the charge of an electron.

Putting these together yields:

$$\text{Edissipated} = \text{ned}^2/\text{mt}$$

If we assume that the distance traveled by the charge packet as it moves from one bucket to the next is.1 microns (10^{-5} centimeters), that one packet has 100 electrons, that the mobility is 1800 cm^2/Vs (the approximate mobility of diamond or silicon near room temperature), that the time allotted for one operation is 1 nanosecond, and the charge e on an electron is 1.6×10^{-19} coulombs, then the energy dissipated is approximately 10^{-21} joules.

This estimate is based on a classical model and involves some significant simplifications. Most significantly, the mean free path will of necessity be less than.1 micron in a device whose maximum dimension is.1 micron, and so the meaning of the electron mobility m is somewhat obscured.

A more detailed quantum analysis is essential as the size becomes smaller. A rough approximation to such an analysis can be based on the observation that the resistance of the channel of a sufficiently small FET (*e.g.*, one in which the channel width is perhaps a few nanometers and only a single transmission mode exists) is roughly half of h/e^2, where h is Planck's constant.

This is a fundamental unit of resistance of about 26K ohms. A packet of n electrons flowing in time t produces a current of ne/t. The power produced by such a current flow is $I^2 \times R$, while the energy dissipation per logic operation is $I^2 \times R \times t$, or $(ne/t)2 \times h/(2e^2) \times t$ which simplifies to:

$$\text{Edissipated} = \text{hn}^2/2\text{t}$$

For a switching time t of 1 nanosecond and a packet size n of 100 electrons, this is 3×10^{-21} joules for each packet that is passed through the channel of the FET. Under the specific conditions considered here this second approximation agrees at least in order of magnitude with the estimate based on a classical model. It is interesting to note that the two equations have significant differences. The resistance of the channel in the classical case varies with the length of the channel. The length of the channel is less significant in the nonclassical case. The size of the charge packet, n, also influences energy dissipation in different ways in the two cases. In the classical case, the energy dissipation is a function of n rather than n^2.

In the classical case, two electrons drifting under the influence of the same electric field in separate regions of the channel will dissipate twice as much energy as a single electron. In effect, the resistance has been halved because the number of charge carriers has been doubled.

A similar effect would occur in a doped semiconductor if the doping density were doubled. If we assume a narrow channel that can only support a single transmission mode, however, then doubling the size of the charge packet will cause more electrons to move through the same channel in the same time period, with a resulting increase in interactions among the electrons. The reader should note that no claim of universality or high accuracy is implied for either of the above equations. They are rough approximations that apply only under a limited range of conditions.

These conditions need not, in general, be true. In particular, it would seem attractive to design devices in which single electrons are always close to their ground state both during charge transport and during switching operations.

This would occur if the electron were always confined in a time varying potential well of relatively small dimensions. Devices with this property should be feasible when we are able to manufacture devices of sufficiently small size. This should have a favourable influence on energy dissipation. The major factor ignored in this analysis is the energy dissipated by the clock signals. Many methods of providing periodic clocking pulses are possible. Three methods are discussed in the next section.

The mobility can be substantially increased by reducing the temperature. Mobilities greater than 1,000,000 have been demonstrated in GaAs-AlGaAs HFET's at around 10 K. While the impact of this is unclear when the device size is much smaller than the mean free path (which occurs with such high mobilities), it does mean that larger devices could be made which would have very low energy dissipation. The energy dissipation for this type of reversible electronic logic (as estimated here) is greater than the dissipation of Drexler's rod logic operated at a cycle time of 100 picoseconds (some 10 times faster than the 1 nanosecond considered here for the RCT). When RCT device operation using single electrons becomes feasible, then energy dissipation using RCT's would presumably be reduced.

Likharev's parametric quantron has an energy dissipation which (when adjusted for the temperature difference and the differing speeds of operation) is not as good as rod logic. The parametric quantron also requires low temperature. Rod logic should work at room temperature. It is possible that mechanical molecular logic might have very favourable energy dissipation properties at low temperature. While it is premature to draw conclusions it is not obvious that electronic devices must necessarily prove better than molecular mechanical devices in terms of energy dissipation at a given speed and temperature of operation.

CLOCKING

To insure that overall energy dissipation is reduced we need a method of providing suitable clocks that does not itself dissipate too much energy. We first consider a conceptually simple method of providing arbitrarily complex clock signals which can be made to dissipate as little energy as desired.

The loads presented to the clock in the proposals discussed here are purely capacitive. If we assume, for the moment, that the energy dissipated when a clock signal charges and discharges such purely capacitive loads is $1/2\ CV^2$, then the only method of reducing the energy dissipation is to reduce the voltage. Assume, for the moment, that we want a clock signal to increase from 0 volts to 5 volts. If we simply connect the clock line to a 5 volt source, then the energy dissipated will be $1/2\ CV^2 = 12.5$ C joules.

If we have 10 voltage supplies, with voltages of 0.5, 1.0, 1.5, 2.0, 2.5, 3.0, 3.5, 4.0, 4.5, and 5.0 volts, then we can successively connect our clock line to each voltage supply in turn. The voltage to the clock line would then be stepped up in 0.5 volt increments, and so the energy dissipated by a single step would be $1/2\ C(V/10)2 = 0.125$ C (*e.g.*, the energy dissipation per step is decreased by a factor of 100 because the size of the voltage step is decreased by a factor of 10).

To increase the voltage from 0 to 5 volts involves 10 steps, and so we will dissipate $10 \times (0.125\ C)$ or 1.25 C joules. This is 10 times smaller than the 12.5 C joules we would have dissipated using a single 5 volt supply. Thus, be using 10 DC power supplies and switching the clock line successively from one power supply to the next, we have reduced energy dissipation by a factor of 10. Clearly, we can reduce energy dissipation by a factor of N if we use N DC power supplies.

Notice that the larger we make N the more time we must take to increase the voltage from 0 to 5 volts. Each time we take a step in voltage we must wait somewhat longer than one RC time constant for the voltage to settle at the new value. This settling time will remain about the same for each step, even as we decrease the size of the voltage step by increasing N. As N increases and the number of steps increases, the total time taken for N steps will also increase. This approach is very flexible because the clock waveform can rise and fall at arbitrary times. Complex clock signals can be easily generated. There are many alternative clocking schemes. While the method just described is asymptotically non-dissipative if a sufficiently large number of DC power supplies are used, it still involves moving electrons through wires to carry the clock signal throughout the circuit. This does not appear to be the major limiting factor in energy dissipation when using today's planar technology. However, if in the future we use single electron logic devices, then moving large numbers of electrons over long distances to provide the clock signals would seem inappropriate.

There are approaches to clock distribution which avoid the need to move charge through wires. A conceptually simple approach is to provide clock distribution by the use of charges that are fixed to a rotating tube or disk. A facing disk or a smaller concentric drum would carry the logic circuitry. As the disk spun the charges fixed to its surface would move, but this motion would not involve electrons flowing within a conductor. The electric charges on the surface of the disk would then move past the logic elements on the facing surface of the opposing disk, providing the repetitive voltage changes required to clock the logic circuitry. In principle this method of moving charges would result in no conventional resistive losses at all because the charge would not be moving through a conductor.

There would still be radiative losses caused by the acceleration of the charged particles, and electrostatic attraction and repulsion would produce forces between the charges on the two disks which would lead to losses simply by alternately compressing and decompressing the material of the disk. While this approach might not be attractive with current technology, it might prove attractive at some point in the future if strong materials could be rotated at high speeds, thus providing a high interfacial velocity between the two opposing surfaces; and the pattern of charges on the surface could be both precisely controlled and have high resolution.

This should eventually be feasible, though whether or not it will prove competitive with other approaches is uncertain. Another method of providing a time-varying electric field would be to use a simple rotating electric field. Such a rotating electric field could then be used to clock the circuit. While the methods described here do not obviously lend themselves to such an approach, helical logic uses such rotating electric fields to drive thermodynamically reversible logic operations.

Charge is transported along a helix much as water is transported along an Archimedes screw. Single electron versions of helical logic should eventually be feasible. Rotating electric fields can be provided by several methods, including cavity resonators or multiple LC oscillators connected to plates arranged in an appropriate pattern in space.

OTHER RCT'S

The 3-bucket RCT is not the only type of RCT that is useful. For example, if we use two condition buckets C_1 and C_2, and require that the charge packets in C_1 and C_2 be complementary (*e.g.*, C_1 is the negation of C_2); and if we use a single source bucket S and two destination buckets D_1 and D_2; then we can implement the primitive "Transfer charge from S to D_1 if C_1 is false, but from S to D_2 if C_1 is true." We will again need a pre-condition that effectively prevents the merging of two packets; we omit the precise statement of this precondition.

This 5-bucket primitive is logically complete (we leave this as an exercise for the reader) and has certain implementation advantages. In particular, because both C_1 and its complement C_2 are available, we need only "steer" the charge in the source bucket to either of the destination buckets. We need not "bottle up" the charge on the source bucket in the event the transfer is not to take place (as occurs with the 3-bucket RCT). Because the charge packet in bucket S is guaranteed to move to some destination bucket (either D_1 or D_2), the actual voltage that must be generated to prevent the charge from moving in the wrong direction is smaller than the voltage required in the simple RCT to prevent the charge from transferring to its single destination bucket.

The same logic function could also be implemented in a somewhat different fashion by using a 4- bucket RCT with a single source bucket S, a single condition bucket C and two destination buckets D_1 and D_2. This is illustrated in Figure. In this alternative implementation, the charge in C would control a single FET between S and D_1, while S and D_2 would always be connected. The key to the conditional transfer is the timing of the clock lines for the two destinations D_1 and D_2.

In particular, we would transfer charge to D_1 through the FET first. During this transfer, D_2 would be maintained in an unreceptive state (the clock for D_2 would be negative). After transfer of charge to D_1 through the FET, the clock for D_2 would be made positive.

This shift would cause charge to transfer from S to D_2 only if the charge had not previously been transferred to D_1 through the FET. Thus, charge would be moved conditionally to either D_1 or D_2 depending on the condition C. In some sense, this is logically equivalent to the simple 3-bucket RCT operation followed by the transfer of the charge in S to the second destination bucket D_2.

A 4-bucket RCT with a single source bucket S, a single destination bucket D and two condition buckets C_1 and C_2 could be used to transfer charge only if both C_1 and C_2 were at logic 0. Arbitrary switching functions could be imposed between the source and destination buckets, and the number of condition buckets could be increased arbitrarily.

A Planar Version of RCT Logic

It would be advantageous to have a planar layout for an RCT that used only sine waves for clocking. This would be easy to implement in current planar technology, and the clock signals could be generated by LC oscillators. These constraints can be met. In particular, we start by considering a 3-phase CCD as illustrated in figure. The three clocks, F0, F120 and F240, represent three sine waves offset by 0, 120, and 240 degrees respectively. Charge packets will move from left to right as the three clock signals vary sinusoidally over time. The illustration shows a top view, the black squares being the metalization regions. The metalization typically would be on top of SiO2, with the bulk Si beneath that. A profile view of charge transfer is given in figure. To provide a switching operation, we must provide some sort of choice for the charge packets moving along the CCD. This is illustrated in figure, where the source CCD shift register divides into two offspring.

To control the direction in which the charge packets are moved, we assume that two "condition" CCD shift registers provide the needed condition packets that will cause charge to select one offspring shift register over the other. Note that this device is just a 5-bucket RCT with a somewhat different geometry and clocking sequence.

It uses 3-phase sine wave clocks to move charge through the device. An additional fourth phase (F180) is required to control the conditional charge transfer from the single large source well (the large rectangle in the figure) to the two destination wells directly to its right. In this particular planar RCT, Condition1 and Condition2 are two CCD shift registers which hold complementary charges. If a potential well in the Condition1 shift register contains a charge packet, then the corresponding potential well in the Condition2 shift register does not (and vice versa).

The condition charge packet moves from left to right. When it reaches the well clocked by F180, the presence or absence of the charge packet changes the voltage on the corresponding gate electrode. There are two gate electrodes controlled by the two complementary charge packets. The gate electrodes are between the source potential well (the large rectangular metalization) and the two destination potential wells to its right. The gate electrodes determine the destination well into which the charge packet in the source potential well moves.

The voltage of the fourth phase, F180, is chosen so that the potential on the gate electrode is somewhat lower than the potential in the two adjacent potential wells when those two potentials are equal (*e.g.*, when F120 = F240)and when the condition charge packet is absent. Thus, when the condition charge packet is absent, the charge packet in the source potential well will flow through the corresponding gate potential well and then into the corresponding destination potential well. When the condition charge packet is present, then the corresponding gate potential will be increased and the charge packet in the source potential well will be prevented from moving into the corresponding destination potential well.

In this case the charge packet in the source potential well will find there is no barrier preventing it from moving into the other destination potential well. Some clock signals are marked with a prime, as F'. The clock lines marked with a prime are the same as the those that are not so marked except for their DC bias level. Because the condition shift registers and the source/destination shift registers share the metalization for the F180 clock, and because basically different things must occur when a charge packet passes beneath the F180 metalization in the two cases, the DC bias levels for the two sets of clock signals need to be separately adjustable.

Otherwise, when the gate electrode blocked a charge packet from moving from the source to the unselected destination, the same voltages applied to the metalizations on the corresponding condition shift register would likewise prevent the condition charge packet from moving along the condition shift register: this, of course, would cause problems. By adjusting the DC bias on the condition register clocks, we can insure that condition charge packets will move from left to right in a regular manner regardless of the voltage variation on the F180 clock caused by the presence or absence of the condition charge packet.

In the same way, by adjusting the bias on the source/destination clocks, we can insure that the voltage variation on the gate electrodes will prevent charge packets from moving into the wrong destination, thus insuring reliable switching action. In a real circuit, many RCT devices would be interconnected by a complex pattern of CCD shift registers. In most cases, this will force the data in one CCD shift register to cross over other shift registers. The use of a planar geometry makes the design of such a cross-over non-obvious.

A relatively simple approach would be to use two crossing sets of metalizations which define two CCD shift registers, and then apply a bias voltage to inactivate one of the shift registers. A charge packet would be carried along the active shift register and would not be diverted into the inactive shift register because the bias voltage applied to the inactive shift register would eliminate the potential wells into which the charge packet might fall. By alternately inactivating one or the other shift register, charge packets could be carried across in first one and then the other direction.

This can be likened to a traffic crossing with a traffic light, where two cars approaching at right angles cannot simultaneously occupy the intersection. The traffic light is first green in one direction, allowing one car to cross; and is then green in the other direction, allowing the other car to cross.

To summarize this planar RCT: a charge packet moving along a CCD shift register will be moved into one of two "offspring" or "descendant" shift registers. Which descendant is selected will depend on which of two condition shift registers is occupied by a charge packet. This operation is logically universal: it can be used to implement a Fredkin gate and therefore any reversible logic circuit. Just as this planar RCT is conceptually similar to a 5-bucket RCT, so too a planar version similar to the 4-bucket RCT would also be feasible and would have the practical advantage that only a single condition would be required to operate it.

The planar 5-bucket RCT suggested here is intended to show that planar RCT's using simple sine waves for clocking are feasible. The design of an actual device might well differ significantly.

NANOROBOT DEXTERITY AND MOBILITY

Manipulation and mobility are crucial basic capabilities in most classes of medical nanodevices. Manipulation includes handling fluids, biological objects such as tissue matrix fibres or cellular elements, and nanomachines or their components. Physicians must be able to direct tissue or cell-repair nanorobots to travel to a specific site where treatment is required, and once there, to manipulate the local environment to achieve the desired results. Nanodevice mobility in vivo makes possible the rapid reconfiguration of nanomedical communication, navigation, and power systems, while ex vivo mobility allows the design of more robust diagnostic and personal defensive systems. In vivo locomotion also permits precise mapping of the internal regions of the human body across many size and time scales, both for diagnostic and for therapeutic purposes.

Adhesion and Fluid Transport

An understanding of surface forces is an essential preliminary to any study of mechanisms of manipulation or locomotion. Such adhesive forces become important at the submicron scale where nanorobotic parts are meshing with and sliding against one another. Surface adhesion forces may cause tools, parts or workpieces to stick together. Surface forces may cause airborne or solvent-immersed nanodevices to adhere to container walls, to other dry or humid surfaces, or to each other possibly causing clumping.

For a 1-micron object, the adhesive forces may exceed gravitational and inertial forces by a factor of 10^6 or more. Capillary forces also are important in fluid transfers inside nanodevices, in fluid transfers between nanodevices or between nanodevices and their operating environment, and during nanodevice tasks requiring locomotion through, or limb/object manipulation in, a watery environment. Surface tension forces are especially relevant during locomotion missions requiring passage through air/water interfaces, as might occur in the lungs. Bowling classifies adhesive forces between objects into three categories.

The first category includes long-range attractive interactions that may bring a particle to a surface and establish the adhesion contact area, such as van der Waals forces, electrostatic forces, and magnetic forces. A second category of forces are the interfacial reactions that help to define the adhesion area, most importantly the capillary forces arising from the establishment of liquid or solid bridges between particle and surface, but also including other effects such as sintering (diffusion and condensation), diffusive mixing, mutual dissolution and surface alloying, which will not be considered further here.

In the third category are very short-range interactions that may strengthen adhesion after an adhesive contact area has already formed; such forces include chemical bonds of all types and various non-covalent bonds such as hydrogen bonds that have already been described.

NANOMANIPULATORS

Numerous simple actuators have been developed for microelectro-mechanical systems or MEMS. A few of these designs might possibly be useful in early nanomedical systems. All major components of an articulated micromanipulator having multiple degrees of freedom, workspaces on the order of 1 mm^3, positional resolution up to ~10 nm, operating frequencies up to 10 KHz, and slewing speeds up to 4 mm/sec have been demonstrated experimentally; complete systems have even been proposed. Actuation and manipulation may be very broadly defined. For example, self-assembling and self-disassembling ~300 micron gold foil cubes have been fabricated and cycled repeatedly at ~1 Hz; "silicon origami" has also been described. non-mechanical actuation has also been explored.

In the Magnetic Stereotaxis System, a helmet with a cubic array of six superconducting coils is used to apply a 0.2 N force to a 3.2-mm wide, 4.7-mm long permanently magnetized cylindrical pellet that is embedded in brain material, causing stepped movement of the pellet through the brain material to a specified location with ~1 mm placement accuracy without any direct physical contact.

However, in the usual definition of a manipulator, an ergomechanical transducer converts chemical, electrical, mechanical, acoustical, or some other form of energy into the mechanical energy of a manipulatory device which then exerts useful forces on objects or materials in the environment. In most cases a force-transmitting physical structure is required which may include both heavy load-bearing elements such as girders, struts, pneumatic tubes or rotary joints, and fine control elements such as stress cables, valves or switches.

An end-effector (*e.g.*, gripper, cutter or molecular jig) is usually employed to focus and redirect the application of force at the distal end of the manipulation device, or to perform specialized tasks. Sensors permit feedback control of manipulator motions as well as verification that the assigned manipulatory task has been properly executed.

Large numbers of manipulators can form ciliary arrays, allowing convenient mass transport of materials; other classes of manipulators may be useful in bulk disassembly of materials. The diverse concepts outlined in this Section demonstrate the enormous range of possible nanomanipulator designs. These concepts are not intended as specific engineering proposals but rather as illustrations of certain useful classes of manipulators representing alternative design pathways that could provide the desired capability.

No effort has been made to produce optimal systems that minimize energy dissipation, maximize speed, minimize parts count, or simplify manufacture.

BIOLOGICAL CILIA

Biological cilia are motile, hairlike cylindrical processes typically 200–300 nm in diameter, 2–20 microns in length, and ~10^{-16}–10^{-15} kg in mass. They are found in large numbers on some cell surfaces. Unicellular organisms use cilia both for locomotion and for food collection, while multicellular organisms use cilia primarily for mass transport of the environment past the cell. Each cilium is bounded by an extension of the cellular plasma membrane, rising from a cytoskeletally-anchored centriole-like basal body consisting of an array of 9 triplets of common-walled microtubules ~150 nm in diameter and 300–500 nm in length. Sprouting upward from the basal body is the primary ciliary structure, called the axoneme.

The axoneme is a cylinder of microtubules with 9 outer doublets of microtubules and two additional complete microtubules in the centre, often called the central pair. The 9 outer doublets are extensions of two of the three microtubules comprising each of the nine basal body triplets. Each outer doublet of the axoneme consists of the A tubule (a complete 25-nm outer-diameter microtubule with the usual 13 protofilaments made of tubulin dimers), the B tubule (an incomplete microtubule with only 10–11 protofilaments), and a common wall containing the protein tektin (an intermediate-filament-like protein with high tensile strength).

RISK ASSESSMENT OF NANOPARTICLES

In biotechnology, different types of nanoparticleshave found extensive use in the cosmetics industry and medicine. In biomedicine, they are used as tools such as imaging agents and drug carriers. For example, liposomes are commonly applied nanoparticles used for delivery of therapeutic drugs, vaccines or other molecules.

Another fast developing trend in biomedicine, therapeutic use of siRNA, has been combined with nanotechnology for treatment of cancer. Use of ananoparticle/siRNA delivery system prevents the RNA molecules from being destroyed by the body before reaching their target. For example, a delivery system in which nanoparticles are used to transport siRNA into cells has been reported for treatment of Ewing's Sarcoma.

The siRNA molecules target growth-promoting genes in sarcoma tumors and use of this delivery system was shown to reduce cell replication in mice grafted with human tumors, by up to 80 per cent. Because of their ultra-small size, nanoparticles can penetrate cell membranes and integrate themselves into larger molecules. They can resist cellular defence systems but are large enough to interfere with cell processes. Despite widespread use in public consumables such as makeup and creams, and the knowledge that very traits that make them useful might also render them toxic, thorough testing on the safety of nanoparticles, once absorbed through the skin, has not been done.

When used for remediation, their release in the environment is also risky due to possible exposure to humans and other animal species. In a press release on April 17, 2007, the American Association for Cancer Research (AACR)reported research presented at the 2007 annual meeting that suggests nanoparticles could cause cancer and should be thoroughly investigated and used with caution. Researchers at the University of Massachusetts studied two types of nanoparticles (silica and C60 fullerene), in MCF-7 and breast cancer cells, and found an increase in DNA damage (single and double-stranded breakages) with both dose- and time-dependent results.

Although DNA breakages do not necessarily mean a substance is cancer-causing, it is widely accepted that chemicals causing DNA damage are highly likely to promote mutations which can lead to cancer. Nanotechnology is expected to be the basis of many of the main technological innovations of the 21st century. Research and development in this field is growing rapidly throughout the world.

A major output of this activity is the development of new materials in the nanometre scale, including nanoparticles. These are usually defined as particulate materials with at least one dimension of less than 100 nanometres (nm). One nanometre is 10^{-9} m. By comparison, a human hair is approximately 70,000 nm in diametre, a red blood cell is approximately 5,000 nm wide and simple organic molecules have sizes ranging from 0.5 to 5 nm

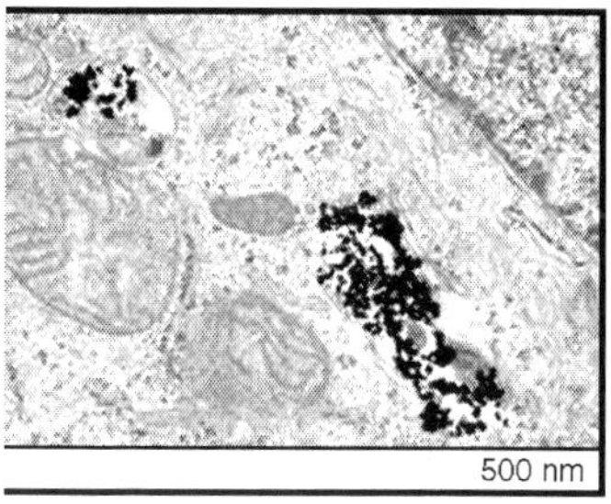

Nanoparticles include carbon nanotubes, metal nanowires, semiconductor quantum dots and other nanoparticles produced from a huge variety of substances. Responsible development of any new materials requires that risks to health and the general environment associated with the development, production, use and disposal of these materials are addressed.

This is necessary to protect workers involved in production and use of these materials, the public and the ecosystem. However, it also helps inform the public debate about the development of these new, potentially beneficial, materials.

Assessment of health risks arising from exposure to chemicals or other substances, requires understanding of the intrinsic toxicity of the substance, the levels of exposure (by inhalation, by ingestion or through the skin) that may occur and any relationship between exposure and health effects. Concerns about the lack of knowledge and possible risks arising from exposure to nanoparticles led the UK Government to request advice from the Royal Society and Royal Academy of Engineering and to the formation of their Nanoscience and Nanotechnology Working Group. The IOM has unique and extensive experience concerning the potential risks from particles and fibres. Our approach combines expertise and state-of-art techniques to identify and characterise exposure scenarios, conduct toxicity evaluations and undertake risk assessments to identify, characterise and estimate the relative safety of nanoparticles. This approach allows us to extend current knowledge and facilitate comparisons between existing substances and new nanoparticles.

IOM nanoparticle activities and initiatives:

- The Safety of Nano-materials Interdisciplinary Research Centre (SnIRC) Initiative:
- IOM has formed a multidisciplinary consortium of internationally recognised researchers with unique and extensive experience to address nanoparticle risk issues. The SnIRC initiative is based on a long-standing collaboration between the IOM and the Universities of Edinburgh, Aberdeen and Napier. Members of the group include Dr Rob Aitken, Prof Ken Donaldson, Prof Jon Ayres, Prof Vicki Stone and Dr Lang Tran.

Current and recent IOM work:

- *Emergnano*: A review of completed and near completed environment, health and safety research on nanomaterials and nanotechnology Defra Project,
- The Enrhes project was an EU FP7 project which performed a comprehensive and critical scientific review of the health and environmental safety of fullerenes, carbon nanotubes (CNTs), metal and metal oxide nanomaterials. The review considers sources, pathways of exposure, the health and environmental outcomes of concern, in the context illustrating the state-of-the-art in the field and informing the regulation of the potential risks of engineered nanoparticles. Aitken RJ, Aschberger K, Baun A, Christensen F, Fernandes T, Foss Hansen S, Hankin SM, Hartmann NB, Hutchison G, Johnston H, Micheletti C, Peters SAK, Ross BL, Solkull-Kluettgen B, Stark D, Stone V and Tran L Engineered Nanoparticles: Review of Health and Environmental Safety.
- Particle_Risk was a collaborative research project funded by the European Commission FP6-NEST programme. It involved 7 research organisations throughout Europe and is led by IOM (Project leader Lang Tran). The aim of Particle_Risk Project was to acquire a bank of NESTP (Fullerene C60, Carbon Nanotubes, Carbon Black, CdTe Quantum dot and nano sized gold) and to assess the health risk from exposure to these materials through air or the food supply with a work programme, integrating in vitroexperiments, animal models of healthy/susceptible individuals and exposure/risk assessment.
- Tran CL, Hankin SM, Ross B, Aitken RJ, Jones AD, Donaldson K, Stone V, Tantra R. An outline scoping study to determine whether

high aspect ratio nanoparticles (HARN) should raise the same concerns as do asbestos fibres.

In the past year, 6 individuals who used a German household cleaner supposedly containing nanoparticles were hospitalized with respiratory problems. However, a later report by the Federal Institute for Risk Assessment (BfR) in Berlin stated that the product did not even contain any nanoparticles.

Despite this report, the damage was done and, although the real cause of their illness was not determined, this event triggered widespread public concern over the safety of products that use nanotechnology. Some even call for a moratorium on the production and sale of nanoparticles until more is known.

The US EPA does not currently regulate nanoparticles but agrees there should be a review of certain compounds and is working on a proposal for standards. Despite the unknowns, the general consensus seems to be that there is a lot of potential in nanotechnology and applications using nanoparticles. However, the bioethics of nanoparticle use in medicine are still in question, and much more research is needed before we can determine how they will react in the human body, and interact with biomolecules, and how to ensure their safe use.

CHEMICAL PROCESSING AND SYNTHESIS

The chemical processing and synthesis of high performance technological components for the private, industrial, and military sectors requires the use of high purity ceramics, polymers, glass-ceramics and material composites. In condensed bodies formed from fine powders, the irregular particle sizes and shapes in a typical powder often lead to non-uniform packing morphologies that result in packing density variations in the powder compact.

Uncontrolled agglomeration of powders due to attractive van der Waals forces can also give rise to in microstructural inhomogeneities. Differential stresses that develop as a result of non-uniform drying shrinkage are directly related to the rate at which the solvent can be removed, and thus highly dependent upon the distribution of porosity. Such stresses have been associated with a plastic-to-brittle transition in consolidated bodies, and can yield to crack propagation in the unfired body if not relieved.

In addition, any fluctuations in packing density in the compact as it is prepared for the kiln are often amplified during the sintering process, yielding inhomogeneous densification. Some pores and other structural defects associated with density variations have been shown to play a detrimental role in the sintering process by growing and thus limiting end-point densities. Differential stresses arising from inhomogeneous densification have also been shown to result in the propagation of internal cracks, thus becoming the strength-controlling flaws. Inert gas evaporation and inert gas deposition are free many of these defects due to the distillation (c.f. purification) nature of the process and having enough time to form single crystal particles, however even their non-aggreated deposits have lognormal size distribution, which is typical with nanoparticles. The reason why modern gas evaporation techniques can produce a relatively narrow size distribution is that aggregation can be avoided. However, even in this case, random residence times in the growth zone, due to the combination of drift and diffusion, result in a size distribution appearing lognormal.

It would therefore appear desirable to process a material in such a way that it is physically uniform with regard to the distribution of components and porosity, rather than using particle size distributions which will maximize the green density. The containment of a uniformly dispersed assembly of strongly interacting particles in suspension requires total control over interparticle forces. Monodisperse nanoparticles and colloids provide this potential.

Monodisperse powders of colloidal silica, for example, may therefore be stabilized sufficiently to ensure a high degree of order in the colloidal crystal or polycrystalline colloidal solid which results from aggregation. The degree of order appears to be limited by the time and space allowed for longer-range correlations to be established. Such defective polycrystalline colloidal structures would appear to be the basic elements of submicrometer colloidal materials science, and, therefore, provide the first step in developing a more rigorous understanding of the mechanisms involved in microstructural evolution in high performance materials and components.

SYNTHESIS OF CARBON NANOTUBES

A range of methodologies have been developed to produce nanotubes in sizeable quantities, including arc discharge, laser ablation, high pressure carbon monoxide (HiPco), and vapour liquid solid (VLS) growth. All these processes take place in vacuum or at low pressure with a process gases, although VLS growth can take place at atmospheric pressure.

Large quantities of nanotubes can be synthesized by these methods; advances in catalysis and continuous growth processes are making SWNTs more commercially viable. The first observation of nanotubes was in the carbon soot formed during the arc discharge production of fullerenes. The high temperatures caused by the discharge caused the carbon contained in the negative electrode to sublime and the CNTs are deposited on the opposing electrode. Tubes produced by this method were initially multi walled tubes (MWNTs). However, with the addition of cobalt to the vapourized carbon, it is possible to grow single walled nanotubes.

This method it produces a mixture of components, and requires further purification to separate the CNTs from the soot and the residual catalytic metals. Producing CNTs in high yield depends on the uniformity of the plasma arc, and the temperature of the deposit forming on the carbon electrode. Higher yield and purity of SWNTs may be prepared by the use of a dual-pulsed laser. SWNTs can be grown in a 50 per cent yield through direct vapourization of a Co/Ni doped graphite rod with a high-powered laser in a tube furnace operating at 1200 °C. The material produced by this method appears as a mat of "ropes", 10 - 20 nm in diametre and up to 100 ìm or more in length. Each rope consists of a bundle of SWNTs, aligned along a common axis. By varying the process parametres such as catalyst composition and the growth temperature, the average nanotube diametre and size distribution can be varied. Although arc-discharge and laser vapourization are currently the principal methods for obtaining small quantities of high quality SWNTs, both methods suffer from drawbacks.

The first is that they involve evapourating the carbon source, making scale-up on an industrial level difficult and energetically expensive. The second issue relates to the fact that vapourization methods grow SWNTs in highly tangled forms, mixed with unwanted forms of carbon and/or metal species. The SWNTs thus produced are difficult to purify, manipulate, and assemble for building nanotube-device architectures for practical applications.

In order to overcome some of the difficulties of these high-energy processes, the chemical catalysis method was developed in which a hydrocarbon feedstock is used in combination with a metal catalyst. The catalyst is typically, but not limited to iron, colbalt, or iron/molybdenum, it is heated under reducing conditions in the presence of a suitable carbon feedstock, *e.g.*, ethylene.

This method can be used for both SWNTs and MWNTs; the formation of each is controlled by the identity of the catalyst and the reaction conditions. A convenient laboratory scale apparatus is available from Nanotech Innovations, Inc., for the synthesis of highly uniform, consistent, research sample that uses pre-weighed catalyst/carbon source ampoules.

This system, allows for 200 mg samples of MWNTs to be prepared for research and testing. The use of CO as a feedstock, in place of a hydrocarbon, led to the development of the high-pressure carbon monoxide (HiPco) procedure for SWNT synthesis. By this method, it is possible to produce gram quantities of SWNTs, unfortunately, efforts to scale beyond that have not met with complete success.

Initially developed for small-scale investigations of catalyst activity, vapour liquid solid (VLS) growth of nanotubes has been highly studied, and now shows promise for large-scale production of nanotubes. Recent approaches have involved the use of well-defined nanoparticle or molecular precursors and many different transition metals have been employed, but iron, nickel, and cobalt remain to be the focus of most research.

The nanotubes grow at the sites of the metal catalyst; the carbon-containing gas is broken apart at the surface of the catalyst particle, and the carbon is transported to the edges of the particle, where it forms the nanotube. The length of the tube grown in surface supported catalyst VLS systems appears to be dependent on the orientation of the growing tube with the surface. By properly adjusting the surface concentration and aggregation of the catalyst particles it is possible to synthesize vertically aligned carbon nanotubes, *i.e.*, as a carpet perpendicular to the substrate. Of the various means for nanotube synthesis, the chemical processes show the greatest promise for industrial scale deposition in terms of its price/unit ratio.

There are additional advantages to the VLS growth, which unlike the other methods is capable of growing nanotubes directly on a desired substrate. The growth sites are controllable by careful deposition of the catalyst. Additionally, no other growth methods have been developed to produce vertically aligned SWNTs.

CHEMICAL FUNCTIONALIZATION OF CARBON NANOTUBES

The limitation of using carbon nanotubes in any practical applications has been its solubility; for example SWNTs have little to no solubility in most solvent due to the aggregation of the tubes. Aggregation/roping of nanotubes occurs as a result of the high van der Waals binding energy of *ca.* 500 eV per mm of tube contact. The van der Waals force between the tubes is so great, that it take tremendous energy to pry them apart, making it very to make combination of nanotubes with other materials such as in composite applications.

The functionalization of nanotubes, *i.e.*, the attachment of "chemical functional groups" provides the path to overcome these barriers. Functionalization can improve solubility as well as processibility, and has been used to align the properties of nanotubes to those of other materials. The clearest example of this is the ability to solubilize nanotubes in a variety of solvents, including water. It is important when discussing functionalization that a distinction is made between covalent and non-covalent functionalization.

Current methods for solubilizing nanotubes without covalent functionalization include highly aromatic solvents, super acids, polymers, or surfactants. Non-covalent "functionalization" is generally on the concept of supramolecular interactions between the SWNT and some macromolecule as a result of various adsorption forces, such as van der Waals' and n-stacking interactions. The chemical speciation of the nanotube itself is not altered as a result of the interaction. In contrast, covalent functionalization relies on the chemical reaction at either the sidewall or end of the SWNT.

As may be expected the high aspect ratio of nanotubes means that sidewall functionalization is much more important than the functionalization of the cap. Direct covalent sidewall functionalization is associated with a change of hybridization from sp^2 to sp^2 and a simultaneous loss of conjugation. An alternative approach to covalent functionalization involves the reaction of defects present (or generated) in the structure of the nanotube.

Defect sites can be the open ends and holes in the sidewalls, and pentagon and heptagon irregularities in the hexagon graphene framework (often associated with bends in the tubes). All these functionalizations are exohedral derivatizations. Taking the hollow structure of nanotubes into consideration, endohedral functionalization of SWNTs is possible, *i.e.*, the filling of the tubes with atoms or small molecules. It is important to note that covalent functionalization methods have one problem in common: extensive covalent functionalization modifies SWNT properties by disrupting the continuous n-system of SWNTs

Various applications of nanotubes require different, specific modification to achieve desirable physical and chemical properties of nanotubes. In this regard, covalent functionalization provides a higher degree of fine-tuning the chemistry and physics of SWNTs than non-covalent functionalization. Until now, a variety of methods have been used to achieve the functionalization of nanotubes.

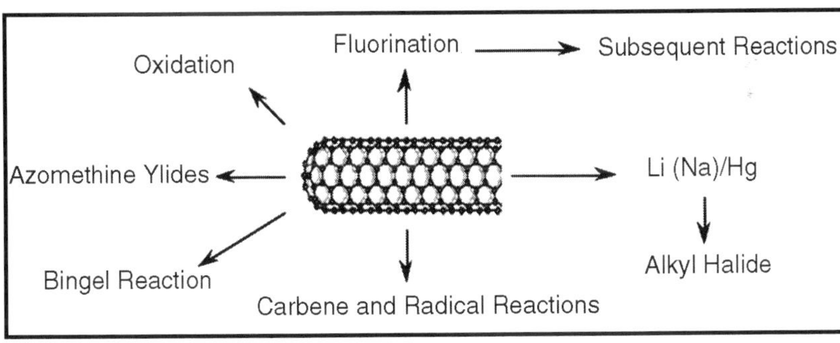

Fig. Schematic Description of Various Covalent Functionalization Strategies for SWNTs.

Taking chemistry developed for C_{60}, SWNTs may be functionalized using 1,3 dipolar addition of azomethine ylides. The functionalized SWNTs are soluble in most common organic solvents. The azomethine ylide functionalization method was also used for the purification of SWNTs. Under electrochemical conditions, aryl diazonium salts react with SWNTs to achieve functionalized SWNTs, alternatively the diazonium ions may be generated *in-situ* from the corresponding aniline, while a solvent free reaction provides the best chance for large-scale functionalization this way. In each of these methods it is possible to control the amount of functionalization on the tube by varying reaction times and the reagents used; functionalization as high as 1 group per every 10 - 25 carbon atoms is possible.

Organic functionalization through the use of alkyl halides, a radical pathway, on tubes treated with lithium in liquid ammonia offers a simple and flexible route to a range of functional groups. In this reaction, functionalization occurs on every 17 carbons. Most success has been found when the tubes are dodecylated. These tubes are soluble in chloroform, DMF, and THF.

The addition of oxygen moieties to SWNT sidewalls can be achieved by treatment with acid or wet air oxidation, and ozonolysis. The direct epoxidation of SWNTs may be accomplished by the direct reaction with a peroxide reagent, or catalytically. Catalytic de-epoxidation allows for the quantitative analysis of sidewall epoxide and led to the surprising result that previously assumed "pure" SWNTs actually contain *ca.* 1 oxygen per 250 carbon atoms.

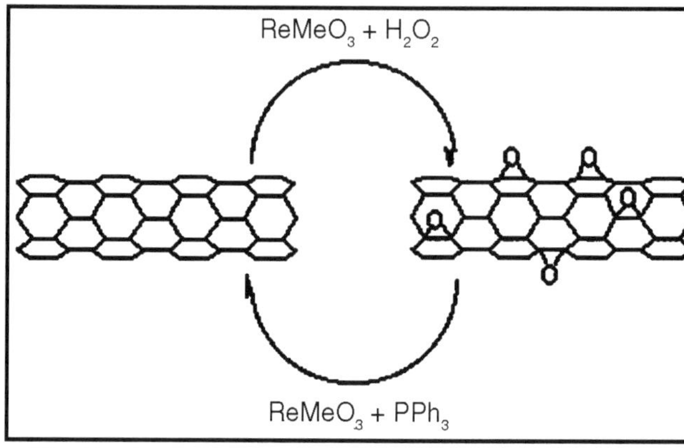

Fig. Catalytic oxidation and de-epoxidation of SWNTs.

One of the easiest functionalization routes, and a useful synthon for subsequent conversions, is the fluorination of SWNTs, using elemental fluorine. Importantly, a C:F ratios of up to 2:1 can be achieved without disruption of the tubular structure. The fluorinated SWNTs (F-SWNTs) proved to be much more soluble than pristine SWNTs in alcohols (1 mg/mL in *iso*-propanol), DMF and other selected organic solvents.

Scanning tunneling microscopy (STM) revealed that the fluorine formed bands of approximately 20 nm, while calculations using DFT revealed 1,2 addition is more energetically preferable than 1,4 addition, which has been confirmed by solid state ^{13}C NMR. F-SWNTs make highly flexible synthons and subsequent elaboration has been performed with organo lithium, Grignard reagents, and amines.

Functionalized nanotubes can be characterized by a variety of techniques, such as atomic force microscopy (AFM), transmission electron microscopy (TEM), UV-vis spectroscopy, and Raman spectroscopy.

Changes in the Raman spectrum of a nanotube sample can indicate if functionalization has occurred. Pristine tubes exhibit two distinct bands. They are the radial breathing mode (230 cm^{-1}) and the tangential mode (1590 cm^{-1}). When functionalized, a new band, called the disorder band, appears at *ca.* 1350 cm^{-1}. This band is attributed to sp^3-hybridized carbons in the tube. Unfortunately, while the presence of a significant D mode is consistent with sidewall functionalization and the relative intensity of D (disorder) mode versus the tangential G mode (1550 – 1600 cm^{-1}) is often used as a measure of the level of substitution. However, it has been shown that Raman is an unreliable method for determination of the extent of functionalization since the relative intensity of the D band is also a function of the substituents distribution as well as concentration. Recent studies suggest that solid state C NMR are possibly the only definitive method of demonstrating covalent attachment of particular functional groups.

TOXICOLOGICAL HAZARDS OF NANOPARTICLES

GENERAL CONCEPTS OF NANOTECHNOLOGY

To use the potential of Nanotechnology in Nanomedicine, full attention is needed to safety and toxicological issues. For pharmaceuticals specific drug delivery formulations may be used to increase the so called therapeutic ratio or index being the margin between the dose needed for clinical efficacy and the dose inducing adverse side effects (toxicity). However, also for these specific formulations a toxicological evaluation is needed. This is particularly true for the applications of nanoparticles for drug delivery. In these applications particles are brought intentionally into the human body and environment, and some of these new applications are envisaged an important improvement of health care. Opinions started to divert when toxicologists claimed that new science, methods and protocols are needed.

However, the need for this is now underlined by several expert reports and more importantly by the following concepts:

- Nanomaterials are developed for their unique (surface) properties in comparison to bulk materials. Since surface is the contact layer with the body tissue, and a crucial determinant of particle response, these unique properties need to be investigated from a toxicological standpoint. When nanoparticles are used for their unique reactive characteristics it may be expected that these same characteristics also have an impact on the toxicity of such particles. Although current tests and procedures in drug and device evaluation may be appropriatc to detect many risks associated with the use of these nanoparticles, it cannot be assumed that these assays will detect all potential risks. So, additional assays may be needed. This may differ depending on the type of particles used, *i.e.*, biological versus non-biological origin.
- Nanoparticles are attributed qualitatively different physico-chemical characteristics from micron-sized particles, which may result in changed body distribution, passage of the blood brain barrier, and triggering of blood coagulation pathways. In view of these charactcristics specific emphasis should be on investigations in (pharmaco)kinetics and distribution studies of nanoparticles. What is currently lacking is a basic understanding of the biological behaviour of nanoparticles in terms of distribution in vivo both at the organ and cellular level.

- Effects of combustion derived nanoparticles in environmentally exposed populations mainly occur in diseased individuals. Typical pre-clinical screening is almost always done in healthy animals and volunteers and risks of particles may therefore be detected at a very late stage.

All would depend on the types of assays used in the preclinical evaluation, which should be considered in the light of the use of the final products. In addition, one cannot rely on the toxicological profile of the bulk material when that material is used in a nanoformulation. What is clear is that the safety evaluation and the risk benefit analysis need to be performed on a case by case basis. The use of nanoparticles as drug carrier may reduce the toxicity of the incorporated drug.

In general the toxicity of the whole formulation is investigated while results of the nanoparticles itself are not described. So, discrimination between drug and nanoparticle toxicity cannot be made. So, there should be a specific emphasis on the toxicity of the "empty" non-drug loaded particles. This is especially important when slowly or non degradable particles are used for drug delivery which may show persistence and accumulation on the site of the drug delivery, eventually resulting in chronic inflammatory reactions.

Evidence for Nanoparticle Toxicity

The toxicity of nanoparticles has originated from inhalation toxicology including the PM_{10} literature (particulate matter with a size below 10 mm), where the 'NP hypothesis' has proved to be a powerful drive for research. An overview of particle terminology in relation to ambient effects. Therefore it relevant to discuss this evidence in the expectation that it will shed light on the toxicity of engineered NPs.

The idea that combustion-derived NPs are an important component that drives the adverse effects of environmental particulate air pollution or PM_{10} comes from several sources:

Table. Various Denominations of Particles in Inhalation Toxicology and Drug Delivery in Relation to their Source (ambient, bulk, engineered)

Particle type	Description
PM_{10}, $PM_{2.5}$	Particle mass fraction in ambient air with a mean diameter of 10 or 2.5 μm respectively. Basis of current standards for ambient particles in Europe and USA
Coarse particles	The mass fraction of PM_{10}, which is bigger than 2.5 μm
Ultrafine particles ($PM_{0.1}$)	The fraction of PM_{10} with a size cut-off at 0.1μm. Contains primary particles and agglomerates smaller than 100 nm
PSP	Poorly soluble particles with low specific toxicity. Maybe be fine or ultrafine. Terminology used in relation to bulk synthetic particles.

	Examples TiO_2, carbon blacks, Amorphous silica, Iron oxides (Fe_2O_3), Zinc oxides (ZnO)
CDNP	Combustion derived nanoparticles, such as diesel exhaust particles (DEP)
DEP	Diesel exhaust particles

- Much of the mass of PM_{10} is considered to be nontoxic and so there has arisen the idea that there is a component(s) of PM_{10} that actually drives the pro-inflammatory effects and combustion-derived NP seems a likely candidate.
- Nanoparticles are the dominant particle type by number suggesting that they may be important and their small size means that they have a large surface area per unit mass. Particle toxicology suggests that, for toxic particles generally, more particle surface equals to more toxicity.
- Substantial toxicological data and limited data from epidemiological sources support the contention that NPs in PM_{10} are important drivers of adverse effects.

The unfavourable health effects of particulate matter (PM) are measurable as exacerbations of respiratory disease and deaths as well as hospitalizations and deaths from respiratory and cardiovascular disease. Inflammation is the common factor that binds together these adverse effects and the ability of NPs to cause inflammation can be seen as an important property. It is not clear what effects of NPs have pulmonary inflammation as a prerequisite and what effects could potentially be driven by exposures below those causing inflammation.

There is also the potential for pulmonary inflammation to result in changes in membrane permeability that in turn may impact the potential for particles to distribute beyond the lung. Some NPs may have the extra potential of affecting cardiovascular disease directly. Vascular function was impaired after inhalation of diesel exhaust particles. However, data to date are limited and not all studies of nanoparticles have shown significant translocation from lung to the blood. In some studies translocation has been rather minimal. Understanding clearance kinetics of inhaled ambient air nanoparticles will also be important in understanding their potential for adverse effects.

The current example in particle toxicology is that ultrafine ambient air particles have the potential of affecting cardiovascular disease both indirectly via pulmonary inflammation and directly through particle distribution. Although important, this property of redistribution has yet to be demonstrated for NPs present in real PM_{10}. It should be noted that there are several mechanisms whereby NPs could lead to inflammatory effects, as is the case for larger particles. These mechanisms are either based on the large surface area of particle core or on soluble components released by the NPs. In addition various chemicals including those of biological origin like endotoxin may be adsorbed onto the NP and released. Several toxicological studies support the contention that NPs in PM_{10} could drive inflammatory effects. There are a number of components of PM_{10} that contribute to the mass but have little toxicity – these include salts such as sulfates, chlorides and ammonium salts and nitrates, but also wind-blown or crustal dust. In fact within PM_{10} there are only few components that toxicologists would identify as likely mediators of adverse effects – *i.e.*, particle surfaces, organics, metals and endotoxin (in some PM_{10} samples). In fact, a large surface area, organics and metals are all characteristic of combustion–derived particles and so these have attracted considerable toxicological attention. However, it is difficult to untangle, in a combustion particle sample, the relative roles of surface, organics and metals, although this has been most attempted in vitro. The aggregation of multiple chemical species including biological compounds like endotoxin limits the extrapolation of the results on the toxicological effects of such particles.

Toxicological Effects of Nanoparticles

The NPs exert some very special properties that are very relevant in the further design of toxicity testing of engineered nanomaterials. An overview of most striking effects of (nano) particles that have been observed over the last decades, along with the particle type that have been tested in this response. Several effects are just quantitatively different from fine particles. In this case nanoparticles may cause the same effects as 'traditional' particles (*e.g.*, inflammation, lung cancer) but they may be more potent because of their greater surface area.

Table. Toxicity of Engineered and Combustion (nano) Particles as Illustrated by their most Unique Adverse Effects in Vivo and in Vitro

Description of finding, in vivo	Particle types
NPs cause pulmonary inflammation in the rat	All PSP
Later studies show that inflammation is mediated by surface area dose	SWCNT, MWCNT
NPs cause more lung tumours than fine particles in rat chronic studies. Effect is surface area mediated	PSP only
NPs cause progression of plague formation (ApoE -/- mice)	SWCNT, PM2.5
NPs affect immune response to common allergens	Polystyrene, CB, DEP
NsP can have access to systemic circulation upon inhalation and instillation surface coating	Specific NP, dependent on
Description of finding, in vitro	
NPs cause oxidative stress in vivo and in vitro, by inflammatory action and generation of surface radicals	PSP, NP general, CNT
NPs inhibit macrophage phagocytosis, mobility and killing	CB, TiO_2
NPs cause platelet aggregation	PM, SWCNT, fullerenes, latex-COOH

surface	
NPs exposure adversely affects cardiac function and vascular homeostasis	PM, SWCNT
NPs interfere with Ca-transport and cause increased binding of pro-inflammatory transcription factor NF-kB	CB (< 100 nm), ROFA, PM2.5
NPs can affect mitochondrial function	Ambient NP,
NPs can translocate to the brain from the nose	MnO_2, Au, carbon
NPs do affect rolling in hepatic tissue	CB

However, nanoparticles could also cause new types of effects not previously seen with larger particles (*e.g.*, mitochondrial damage, uptake through olfactory epithelium, platelet aggregation, cardiovascular effects). These effects depicted, clearly need a new way of handling their toxicology. In addition, epidemiological evidence suggests that these effects occur predominantly in subjects that have an impaired health. This finding should be considered in developing toxicological testing models.

Effects on Blood and Cardiovascular System

The ligand coated engineered nanoparticles are being explored and used as agents for molecular imaging or drug delivery tools. This has led to a considerable understanding of particle properties that can affect penetration in tissue without affecting tissue function. Cationic NPs, including gold and polystyrene have been shown to cause hemolysis and blood clotting, while usually anionic particles are quite nontoxic. This conceptual understanding maybe used to prevent potential effects of unintended NP exposure. Similarly, drug loaded nanoparticles have been used to prolong half-life or reduce side-effects and have shown which particle properties need to be modified to allow delivery, while being biocompatible.

On the other hand, one is trying to find explanations for the increased risk of patients with cardiovascular diseases upon exposure to PM and/or traffic. Several toxicological studies have demonstrated that combustion and model NPs can gain access to the blood following inhalation or instillation and can enhance experimental thrombosis but it is not clear whether this was an effect of pulmonary inflammation or particles translocated to the blood. High exposures to DEP by inhalation caused altered heart rate in hypertensive rats interpreted as a direct effect of DEP on the pacemaker activity of the heart.

Inflammation in distal sites has long been associated with destabilization of atheromatous plaques and both instillation and inhalation of PM cause morphological evidence of atheromatous plaque increase and destabilization in rabbits and mice. Ultrafine carbon black instilled into the blood has been reported to induce platelet accumulation in the hepatic microvasculature of healthy mice in association with prothrombotic changes on the endothelial surface of the hepatic microvessels. Recent studies with carbon derived nanomaterials showed that platelet aggregation was induced by both single and multi-wall carbon nanotubes, but not by the C60-fullerenes that are used as building blocks for these CNT. These data show that not all nanomaterials act similar in this test, and that surface area is not the only factor playing a role here. The data also corroborate the earlier concept developed in medicine that mainly cationic species have an effect on blood clotting. Interestingly, this is the first study that allows bridging of data, since also a real life PM10 sample (SRM1648) was included in the test-series. Actually the PM sample showed a lower effect compared to the carbon nanotubes.

Uptake and Effects of Nanoparticles in the Brain

Nanoparticles can get access to the brain by two different mechanisms, *i.e.*, (1) transsynaptic transport after inhalation through the olfactory epithelium, and (2) uptake through the blood-brain barrier. The first pathway has been studied primarily with model particles such as carbon, Au and MnO_2 in experimental inhalation models in rats. The second pathway has been the result of extensive research and particle surface manipulation in drug delivery. The latter studies suggest that the physiological barrier may limit the distribution of some proteins and viral particles after transvascular delivery to the brain, suggesting that the healthy BBB contains defence mechanisms protecting it from blood borne nanoparticle exposure.

When nanoparticles with different surface characteristics were evaluated, neutral nanoparticles and low concentrations of anionic nanoparticles were found to have no effect on BBB integrity, whereas high concentrations of anionic nanoparticles and cationic nanoparticles were toxic for the BBB. Nanoparticles have been shown to induce the production of reactive oxygen species and oxidative stress and this has been confirmed in the brain after inhalation of MnO_2 nanoparticles. Oxidative stress has been implicated in the pathogenesis of neurodegenerative diseases such as Parkinson's and Alzheimer's diseases. Evidence for the involvement of ambient air nanoparticles in these effects is presented by studies in biopsies from city dwellers. Alzheimer's like pathology was demonstrated in brain sections by increased markers of inflammation and AB42-accumulation in frontal cortex and hippocampus in association with the presence of nanoparticles. Also inhalation exposure of BALB/c mice to particulate matter showed activation of pro-inflammatory cytokines in the brain. Whether this is due to the fraction of combustion nanoparticles remains to be investigated.

NANOMATERIALS IN MEDICINE: NEEDS

Even though there is a considerable amount of data on the toxicity of NPs, this data is mainly based on a small panel of NPs (combustion derived NPs, TiO_2, CB) and the assumption that a lot of effects by particulate matter are driven by the ultrafine particle fraction in it. In most studies the nanoparticles were used as a model for ambient air particle toxicity. One of the more general conclusions is that indeed there is a clear tendency for very small (nano) particles to be more toxic than larger particles with the same chemical composition.

For nanoformulations used in drug delivery the focus in most papers is mainly on obtained reduction of toxicity of the incorporated drug, whereas the possible toxicity of the carrier used is not considered. Especially possible residues of such a treatment may harbour potential local and/or systemic toxic responses. Meant for medical applications certain routine assays need to be performed which will detect a number of potential hazards. However, it can be anticipated that not all hazards are at this moment known for the use of nanoparticles.

In a recent report Costigan reviewed the evidence for toxicity of NPs used in health care products. Her conclusions again stressed the limited availability of toxicity data of the NPs in use. However, in NPs for health care products most if not all mechanisms of toxicity could be identified by conventional hazard identification testing as currently required to comply with the regulations for health care products. Costigan identified four possible mechanisms of NP toxicity, being chemical toxicity of one of the constituents with the same mode of action as the bulk chemical, toxicity due to degradation products, toxicity due to endocytosis of the NPs, and membrane lysis due to the NPs possibly via chemical toxicity.

Although hazard identification is the general approach for safety evaluation of health care products, it is recommen-ded to add testing driven by the anticipated application and classification by risk. Some engineered NPs, which get airborne will pose inhalation hazards, while cosmetics with NPs provide dermal exposures.

For parenteral use interactions with blood components, systemic distribution and kinetics are of importance, when engineered NPs are being used as devices to target drugs to specific tissues, to increase their biological half time, or for imaging purposes. Each nanoparticle formulation should be tested on a case by case basis in the requisite ways focusing on their portal of entry. In this respect also the potential adverse (toxic) effects of empty particles should be considered.

In developing testing procedures and protocols a number of basic issues need to be considered:

- Which effects are specific for nanomaterials, and which effects are merely stronger? Nanoparticles may cause the same effects as 'traditional' particles (eg, inflammation, lung cancer) but they may be more potent because of their greater surface area. Nanoparticles could also cause new types of effects not previously seen with larger particles or bulk chemicals.

- Can we extrapolate available data and concepts? The epidemio-logical evidence on ultrafine particles has revealed several effects, mechanisms of action and susceptible groups upon inhalation of ultrafine particles. Whether these concepts can be used for nano-particles released from manufactured nanomaterials is yet unknown.
- Is our current regulation robust enough to handle risks of nanomaterials? We deal with a growing set of materials of which the properties are largely unknown and for which current testing procedures and legislation might produce false negatives and/or false positives. The central question here is whether current testing and classification protocols are appropriate or sufficient. These will detect certain toxic effects as demonstrated by the studies already published. However, it can be anticipated that not all hazards will be detected and additional specific testing may be needed. Nanotechnology also promotes convergence of technologies, and for example similar materials may be applied in the automotive and the life sciences sector. To stimulate production and marketing of safe nanomaterials exchange of data between sectors is recommended. Sharing of information on the toxicity of nanomaterials, will significantly reduce time to market for many products and producers.
- Should we use the precautionary principle in current regulatory testing? The precautionary principle (PP) is a highly debated issue in international politics and was first added in EU environmental regulation in the Maastricht treaty in Article 174 (ex Art 130r). The PP points out that scientific uncertainty is no reason for inaction if there might be strong adverse effects. It has been criticized for being too vague and too arbitrary to form a basis for rational decision-making. On the other hand the PP does not necessarily imply a complete ban of substances but may be applied in steps of decreasing uncertainty or perceived risk. It is typically applied where scientific information is insufficient, inconclusive or uncertain whilst at the same time there are indications that potential effects may occur. Approaches in risk governance range from early explorations by Swiss RE to more recent detailed reviews such by Renn and Roco and the Health Council of the Netherlands. A crucial difference between both reports is that Renn and Roco place the nanotechnology *products themselves* into the risk issue categories rather than the risk issues that accompany the use of these products. This difference may seem trivial, but is important since it is the purpose and application rather than the device itself that connect hazard to exposure, and therefore creates a risk. For example, it fails to distinguish between brain implants for enhancement purposes and those for fighting tremor in

Parkinson patients. This may pave the way to one nanotechnology product compromising or allowing all the others. Furthermore, it can convey the impression that the broader societal implications and accompanying ethical issues of nanotechnologies are as new as the nanotechnology products themselves, or even are still in the future. However, most of them are not.

3

Characterization of Nanomaterials

Nanomaterials have extremely small size which having at least one dimension 100 nm or less. Nanomaterials can be nanoscale in one dimension, two dimensions, or three dimensions. They can exist in single, fused, aggregated or agglomerated forms with spherical, tubular, and irregular shapes. Common types of nanomaterials include nanotubes, dendrimers, quantum dots and fullerenes.

Nanomaterials have applications in the field of nano technology, and displays different physical chemical characteristics from normal chemicals, i.e.:

- Silver nano,
- Carbon nanotube,
- Fullerene,
- Photocatalyst,
- Carbon nano,
- Silica

Siegel, Nanostructured Materials are Classified as Zero Dimensional, One Dimensional, Two Dimensional, Three Dimensional Nanostructures.

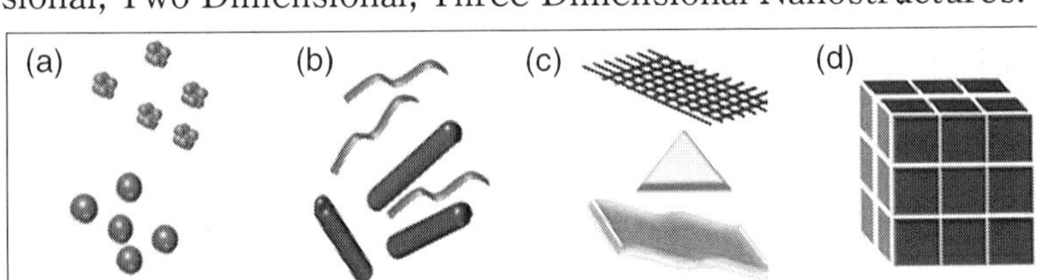

Fig. Classification of Nanomaterials (a) 0D Spheres and Clusters, (b) 1D Nanofibres, wires, and Rods, (c) 2D Films, Plates, and Networks, (d) 3D Nanomaterials.

Nanomaterials are materials which are characterized by an ultra fine grain size or by a dimensionality limited to 50 nm.

Nanomaterials can be created with various modulation dimensionalities as defined by Richard W. Siegel:

- Zero (atomic clusters, filaments and cluster assemblies),
- One (multilayers),
- Two (ultrafine-grained overlayers or buried layers), and
- Three (nanophase materials consisting of equiaxed nanometer sized grains).

IMPORTANT OF NANOMATERIALS

These materials have created a high interest in recent years by virtue of their unusual mechanical, electrical, optical and magnetic properties.

Some examples are:

- Nanophase ceramics are of particular interest because they are more ductile at elevated temperatures as compared to the coarse-grained ceramics.
- Nanostructured semiconductors are known to show various non-linear optical properties. Semiconductor Q-particles also show quantum confinement effects which may lead to special properties, like the luminescence in silicon powders and silicon germanium quantum dots as infrared optoelectronic devices. Nanostructured semiconductors are used as window layers in solar cells.
- Nanosized metallic powders have been used for the production of gas tight materials, dense parts and porous coatings. Cold welding properties combined with the ductility make them suitable for metal-metal bonding especially in the electronic industry.
- Single nanosized magnetic particles are mono-domains and one expects that also in magnetic nanophase materials the grains correspond with domains, while boundaries on the contrary to disordered walls. Very small particles have special atomic structures with discrete electronic states, which give rise to special properties in addition to the superparamagnetism behaviour. Magnetic nano-composites have been used for mechanical force transfer, for high density information storage and magnetic refrigeration.
- Nanostructured metal clusters and colloids of mono-or plurimetallic composition have a special impact in catalytic applications. They may serve as precursors for new type of heterogeneous catalysts and have been shown to offer substantial advantages concerning activity, selectivity and lifetime in chemical transfor-mations and electro-catalysis. Enantioselective catalysis was also achieved using chiral modifiers on the surface of nanoscale metal particles.
- Nanostructured metal-oxide thin films are receiving a growing attention for the realization of gas sensors with enhanced sensitivity and selectivity. Nanostructured metal-oxide finds application for rechargeable batteries for cars or consumer goods. Nanocrystalline silicon films for highly transparent contacts in thin film solar cell and nano-structured titanium oxide porous films for its high transmission and significant surface area enhancement leading to strong absorption in dye sensitized solar cells.

- Polymer based composites with a high content of inorganic particles leading to a high dielectric constant are interesting materials for photonic band gap structure.

MICROSCOPIC STRUCTURE OF NANOMATERIALS

Imagine being able to see the microscopic structure of a material, and noticing that it is composed of many periodic patterns on the order of 100 nm or less. The shape and size properties of those tiny patterns can affect the physical properties of the material that they are part of. For example, solutions of gold nanoparticles that differ only in their size and shape properties at the nanoscale interact with light differently.

The result is different colours. In "nanomaterials," the size and shape of the very small domains that compose the bulk material affect its properties. The dependence of the physical property (such as colour or conductivity) on the size and shape of the nanoscale domains is often a result of *surface*and *interface interactions*.

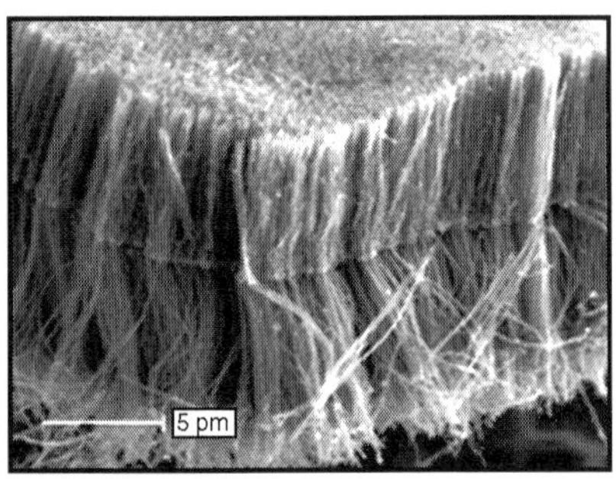

Materials that are comprised of many small structures often have a very large surface area. The reason is that the surface areas of the many small structures together contribute to a very large surface area of the bulk material. For example, if a piece of material the size of a shoe box were to be split up into 2 nm spheres, these spheres would contribute to a surface area of the shoe box that correspond to the area of 10,000 football fields.

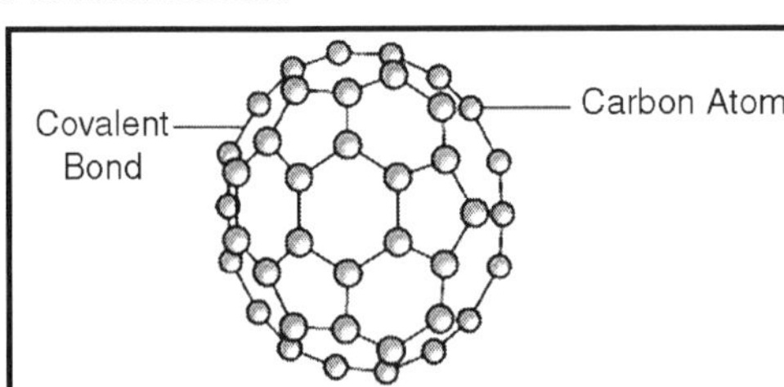

Fig. Sixty Carbon Atoms in the Shape of a Sphere — a Buckball.

It is not surprsing that when the surface area is so large, the physical properties are dominated by surface interactions. A buckyball (short for buckminsterfullerene, a term we explain shortly) is a molecule containing 60 carbon atoms. Each carbon atom is bonded to three adjacent carbon atoms, just as in graphite. However the carbon atoms in a buckyball form a teensy-weensy sphere that's about 1 nanometre in diametre, as shown in Figure. Because one of the properties of carbon atoms is that they can bond to many other types of atoms, researchers can use them to create customised molecules.

DESPERATELY SEEKING BUCKYBALLS

Buckyballs were discovered through an interesting collaboration of researchers from two universities. Richard Smalley at Rice University was studying semiconductor materials. He had a device that shined a laser at a solid sample, vapourized part of it, and analysed the clusters of atoms that formed in the vapour. Meanwhile, at the University of Sussex, Harry Kroto was attempting to reproduce a material found in deep space that generated specific molecular spectra from carbon atoms. Bob Curl, also from Rice University, was doing similar work. Curl ran into Kroto at a symposium and suggested he drop by Rice University because Smalley's scientific instrumentation might be helpful in his work. Kroto dropped by Rice, and after seeing the work that Smalley was doing, he became interested in using that equipment to reproduce his carbon molecules.

Time spent working with high-end scientific equipment is always at a premium, so Kroto had to wait about a year until the equipment was available. In August of 1985, Smalley, Kroto, Curl, and some graduate students performed a series of experiments producing carbon molecules and clusters. They found that under certain conditions, most of the molecules generated contained 60 carbon atoms. Voilà: buckyballs. While many of the atoms in buckyballs are connected together in hexagons (just as in graphite sheets), some of the atoms are connected together in pentagons. The pentagons allow the sheet of carbon atoms to curve into the shape of a sphere. Every buckyball surface contains 12 pentagons and 20 hexagons.

After much discussion and modeling, researchers determined that 60 carbon atoms form a single stable molecule only if they're arranged in 20 hexagons and 12 pentagons that are linked to form a sphere — as it happens, it's the same arrangement of hexagons and pentagons proposed by American architect and engineer Buckminster Fuller for his famed geodesic dome. No wonder these molecules got dubbed buckminsterfullerenes (that's fullerenes for short) in his honour. This type of spherical carbon molecule has been found in various other sizes. The fullerene family of molecules is often identified by the letter C followed by the number of carbon atoms, for example C60, C70, C80.

Creating Buckyballs

Richard Smalley, one of the scientists credited with discovering buckyballs, produced the little wonders in a device by vapourizing carbon with a laser and allowing the carbon atoms to condense. However, this device could only produce a very small number of buckyballs — enough to show their existence but not enough to allow thorough study of their capabilities.

It took collaboration between researchers at the Max Planck Institute in Germany and the University of Arizona to figure out how to make larger quantities of buckyballs. Operating on a shoestring, they vapourized carbon by placing two carbon electrodes close together and generating an electric arc between them in a reaction chamber filled with a low pressure of helium or neon (neither of which will bond with carbon). This method generated much larger quantities of buckyballs than Smalley's device did, although the buckyballs had to be separated from carbon soot by using solvents such as benzene. The method produced enough buckyballs to analyse for their electrical and physical properties. But what if you want to produce buckyballs by the truckload. A little tech school in Massachusetts called MIT, working with a company called Nano-C, Inc., developed a method called combustion synthesis, which produces big enough quantities of buckyballs — at a low enough cost — for use in commercial applications. This method mixes a hydrocarbon with oxygen and burns the hydrocarbon at a low pressure. Nano- C states that its latest version of this process produces buckyball material pure enough (about 95 per cent) to skip the step that separates the product from the carbon soot — and that reduces costs significantly.

CARBON NANOMATERIALS

Although nanomaterials had been known for many years prior to the report of C_{60} the field of nanoscale science was undoubtedly founded upon this seminal discovery. Part of the reason for this explosion in nanochemistry is that while carbon materials range from well-defined nano sized molecules (*i.e.*, C_{60}) to tubes with lengths of hundreds of microns, they do not exhibit the instabilities of other nanomaterials as a result of the very high activation barriers to their structural rearrangement.

As a consequence they are highly stable even in their unfunctionalized forms. Despite this range of carbon nanomaterials possible they exhibit common reaction chemistry: that of organic chemistry. The previously unknown allotrope of carbon: C_{60}, was discovered in 1985, and in 1996, Curl, Kroto, and Smalley were awarded the Nobel Prize in Chemistry for the discovery. The other allotropes of carbon are graphite (sp^2) and diamond (sp^3). C_{60}, commonly known as the "buckyball" or "Buckminsterfullerene", has a spherical shape comprising of highly pyramidalized sp^2 carbon atoms. The C_{60} variant is often compared to the typical soccer football, hence buckyball. However, confusingly, this term is commonly used for higher derivatives. Fullerenes are similar in sheet structure to graphite but they contain pentagonal (or sometimes heptagonal) rings that prevent the sheet from being planar.

The unusual structure of C_{60} led to the introduction of a new class of molecules known as fullerenes, which now constitute the third allotrope of carbon. Fullerenes are commonly defined as "any of a class of closed hollow aromatic carbon compounds that are made up of twelve pentagonal and differing numbers of hexagonal faces." The number of carbon atoms in a fullerene range from C_{60} to C_{70}, C_{76}, and higher. Higher order fullerenes include carbon nanotubes that can be described as fullerenes that have been stretched along a rotational axis to form a tube. As a consequence of differences in the chemistry of fullerenes such as C_{60} and C_{70} as compared to nanotubes, these will be dealt with separately herein. In addition there have also been reports of nanohorns and nanofibres, however, these may be considered as variations on the general theme. It should be noted that fullerenes and nanotubes have been shown to be in flames produced by hydrocarbon combustion. Unfortunately, these naturally occurring varieties can be highly irregular in size and quality, as well as being formed in mixtures, making them unsuitable for both research and industrial applications. Carbon-60 (C_{60}) is probably the most studied individual type of nanomaterial. The spherical shape of C_{60} is constructed from twelve pentagons and twenty hexagons and resembles a soccer ball.

The next stable higher fullerene is C_{70} that is shaped like a rugby or American football. The progression of higher fullerenes continues in the sequence C_{74}, C_{76}, C_{78}, etc. The structural relationship between each involves the addition of six membered rings. Mathematically (and chemically) two principles define the existence of a stable fullerene, *i.e.*, Euler's theorem and isolated pentagon rule (IPR). Euler's theorem states that for the closure of each spherical network, n (n e" 2) hexagons and 12 pentagons are required while the IPR says no two pentagons may be connected directly with each other as destabilization is caused by two adjacent pentagons.

Although fullerenes are composed of sp2 carbons in a similar manner to graphite, fullerenes are soluble in various common organic solvents. Due to their hydrophobic nature, fullerenes are most soluble in CS2 (C60 = 7.9 mg/mL) and toluene (C60 = 2.8 mg/mL). Although fullerenes have a conjugated system, their aromaticity is distinctive from benzene that has all C-C bonds of equal lengths, in fullerenes two distinct classes of bonds exist.

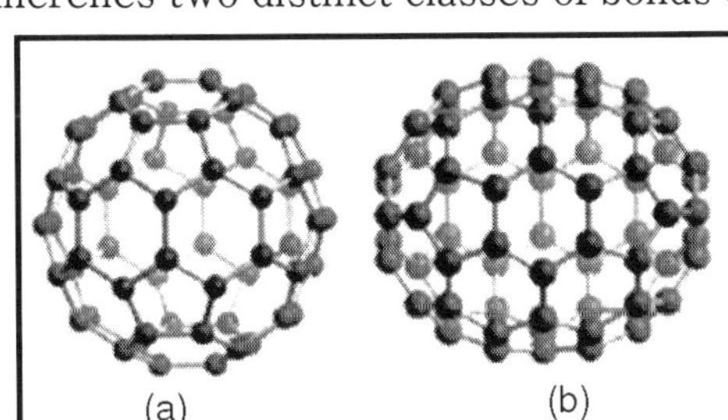

Fig. Molecular Structures of (a) C_{60} and (b) C_{70}.

The shorter bonds are at the junctions of two hexagons ([6, 6] bonds) and the longer bonds at the junctions of a hexagon and a pentagon ([5,6] bonds). This difference in bonding is responsible for some of the observed reactivity of fullerenes.

The first observation of fullerenes was in molecular beam experiments at Rice University. Subsequent studies demonstrated that C_{60} it was relatively easy to produce grams of fullerenes. Although the synthesis is relatively straightforward fullerene purification remains a challenge and determines fullerene's commercial price. The first method of production of measurable quantities of fullerenes used laser vapourization of carbon in an inert atmosphere, but this produced microscopic amounts of fullerenes.

Laboratory scales of fullerene are prepared by the vapourization of carbon rods in a helium atmosphere. Commercial production ordinarily employs a simple ac or dc arc. The fullerenes in the black soot collected are extracted in toluene and purified by liquid chromatography. The magenta C_{60} comes off the column first, followed by the red C_{70}, and other higher fullerenes. Even though the mechanism of a carbon arc differs from that of a resistively heated carbon rod (because it involves a plasma) the He pressure for optimum C_{60} formation is very similar.

A ratio between the mass of fullerenes and the total mass of carbon soot defines fullerene yield. The yields determined by UV-Vis absorption are approximately 40 per cent, 10-15 per cent, and 15 per cent in laser, electric arc, and solar processes. Interestingly, the laser ablation technique has both the highest yield and the lowest productivity and, therefore, a scale-up to a higher power is costly. Thus, fullerene commercial production is a challenging task.

The world's first computer controlled fullerene production plant is now operational at the MER Corporation, who pioneered the first commercial production of fullerene and fullerene products. Endohedral fullerenes are fullerenes that have incorporated in their inner sphere atoms, ions or clusters. Endohedral fullerenes are generally divided into two groups: endohedral metallofullerenes and non-metal doped fullerenes. The first endohedral metallofullerenes was called LaC_{60}. The sign in the name reflects the notion of a small molecule trapped inside a shell. Doping fullerenes with metals takes place in-situ during the fullerene synthesis in an arc reactor or via laser evapouration. A wide range of metals have been encased inside a fullerene, *i.e.*, Sc, Y, La, Ce, Ba, Sr, K, U, Zr, and Hf. Unfortunately, the synthesis of endohedral metallofullerenes is unspecific because in addition a high yield of unfilled fullerenes, compounds with different cage sizes are prepared.

A characteristic of endohedral metallofullerenes is that electrons will transfer from the metal atom to the fullerene cage and that the metal atom takes a position off-centre in the cage. The size of the charge transfer is not always simple to determine, but it is usually between 2 and 3 units but can be as high as 6 electrons. These anionic fullerene cages are very stable molecules and do not have the reactivity associated with ordinary empty fullerenes. This lack of reactivity is utilized in a method to purify endohedral metallofullerenes from empty fullerenes.

The endohedral HeC_{60} and NeC_{60} form when C_{60} is exposed to a pressure of around 3 bar of the appropriate noble gases. Under these conditions it was possible to dope 1 in every 650,000 C_{60} cages with a helium atom. Endohedral complexes with He, Ne, Ar, Kr and Xe as well as numerous adducts of the HeC_{60} compound have also been proven with operating pressures of 3000 bars and incorporation of up to 0.1 of the noble gases.

The isolation of NC_{60}, NC_{70} and PC_{60} is very unusual and unlike the metal derivatives no charge transfer of the pnictide atom in the centre to the carbon atoms of the cage takes place. Although fullerenes have a conjugated aromatic system all the carbons are quaternary (*i.e.*, containing no hydrogen), which results in making many of the characteristic substitution reactions of planar aromatics impossible. Thus, only two types of chemical transformations exist: redox reactions and addition reactions.

Of these, addition reactions have the largest synthetic value. Another remarkable feature of fullerene addition chemistry is the thermodymics of the process. Since the sp2 carbon atoms in a fullerene are paramidalized there is significant strain energy. For example, the strain energy in C60 is ca 8 kcal/mol, which is 80 per cent of its heat of formation. So the relief of this strain energy leading to sp3 hybridized C atoms is the major driving force for addition reactions. As a consequence, most additions to fullerenes are exothermic reactions.

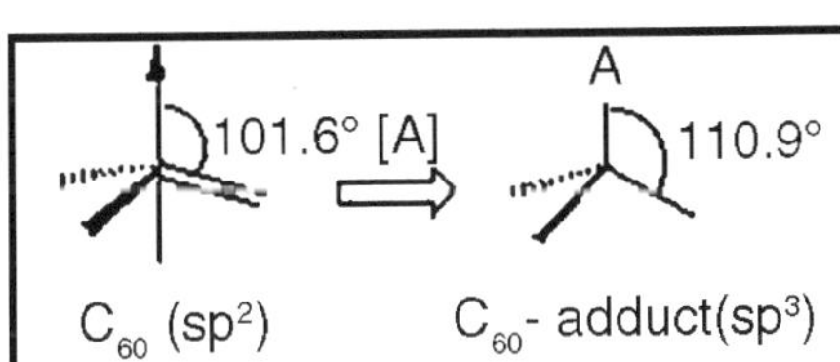

Fig. 2 Strain Release after Addition of Reagent A to a Pyramidalize Carbon of C_{60}.

Cyclic voltammetry (CV) studies show that C_{60} can be reduced and oxidized reversibly up to 6 electrons with one-electron transfer processes. Fulleride anions can be generated by electrochemical method and then be used to synthesize covalent organofullerene derivatives. Alkali metals can chemically reduce fullerene in solution and solid state to form M_xC_{60} (x = 3 – 6).

C_{60} can also be reduced by less electropositive metals like mercury to form C_{60}^{-} and C_{60}^{2-}. In addition, salts can also be synthesized with organic molecules, for example possesses interesting electronic and magnetic behaviour. Geometric and electronic analysis predicted that fullerene behaves live an electro-poor conjugated polyolefin. Indeed C_{60} and C_{70} undergo a range of nucleophilic reactions with carbon, nitrogen, phosphorous and oxygen nucleophiles.

C_{60} reacts readily with organolithium and Grignard compounds to form alkyl, phenyl or alkanyl fullerenes. Possibly the most widely used additions to fullerene is the Bingel reaction, where a carbon nucleophile, generated by deprotonation of á-halo malonate esters or ketones, is added to form a cyclopropanation product. The á-halo esters and ketones can also be generated in situ with I_2 or CBr_4 and a weak base as 1,8-diazabicyclo[5.4.0]unde-7ene (DBU). The Bingel reaction is considered one of the most versatile and efficient methods to functionalize C_{60}.

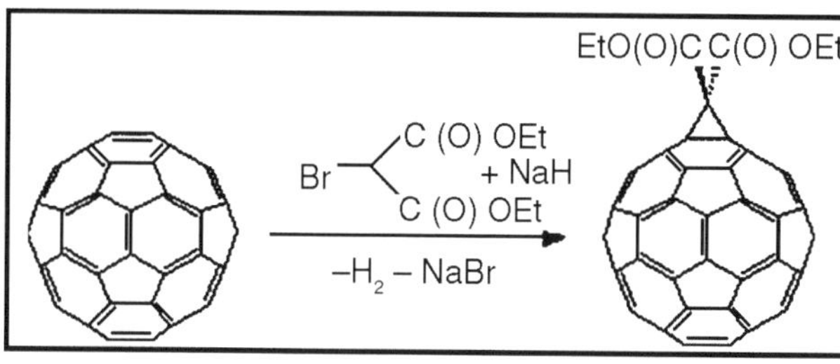

Fig. Bingel Raction of C_{60} with 2-bromoethylmalonate.

Cycloaddition is another powerful tool to functionalize fullerenes, in particular because of its selectivity with the 6,6 bonds, limiting the possible isomers.

The dienophilic feature of the [6,6] double bonds of C_{60} enables the molecule to undergo various cycloaddition reactions in which the monoadducts can be generated in high yields.

The best studies cycloadditon reactions of fullerene are [3+2] additions with diazoderivatives and azomethine ylides (Prato reactions). In this reaction, azomethine ylides can be generated *in situ* from condensation of á-amino acids with aldehydes or ketones, which produce 1,3 dipoles to further react with C_{60} in good yields. Hundreds of useful building blocks have been generated by those two methods. The Prato reactions have also been successfully applied to carbon nanotubes.

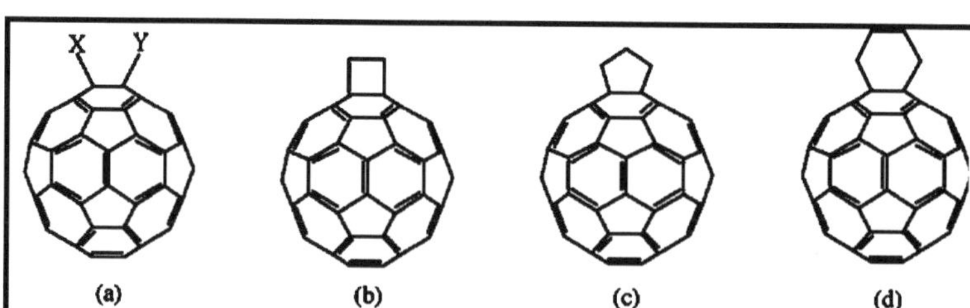

Fig. Geometrical Sapes Bilt onto a [6,6] Rng Jnction:a) Open,b) Four-membered ring,c) Five-membered ring, andd) Six-membered ring.

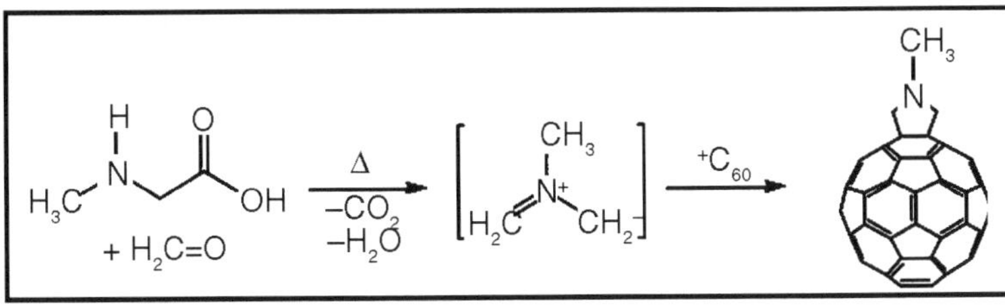

Fig. Prato reaction of C_{60} with N-methyglycine and Paraformaldehyde.

The oxidation of fullerenes, such as C_{60}, has been of increasing interest with regard to applications in photoelectric devices, biological systems, and possible remediation of fullerenes. The oxidation of C_{60} to $C_{60}O_n$ (n = 1, 2) may be accomplished by photooxidation, ozonolysis, and epoxidation. With each of these methods, there is a limit to the isolable oxygenated product, $C_{60}O_n$ with n < 3. Highly oxygenated fullerenes, $C_{60}O_n$ with 3 d" n d" 9, have been prepared by the catalytic oxidation of C_{60} with $ReMeO_3/H_2O_2$.

SELECTED APPLICATION OF NANOMATERIALS

Nanomaterials having wide range of applications in the field of electronics, fuel cells, batteries, agriculture, food industry, and medicines, etc... It is evident that nanomaterials split their conventional counterparts because of their superior chemical, physical, and mechanical properties and of their exceptional formability.

ELECTROCHEMICAL ENERGY CONVERSION DEVICE

A fuel cell is an electrochemical energy conversion device that converts the chemical energy from fuel (on the anode side) and oxidant (on the cathode side) directly into electricity. The heart of fuel cell is the electrodes. The performance of a fuel cell electrode can be optimized in two ways; by improving the physical structure and by using more active electro catalyst. A good structure of electrode must provide ample surface area, provide maximum contact of catalyst, reactant gas and electrolyte, facilitate gas transport and provide good electronic conductance. In this fashion the structure should be able to minimize losses.

Carbon Nanotubes: Microbial Fuel Cell

Microbial fuel cell is a device in which bacteria consume water-soluble waste such as sugar, starch and alcohols and produces electricity plus clean water. This technology will make it possible to generate electricity while treating domestic or industrial wastewater. Microbial fuel cell can turn different carbohydrates and complex substrates present in wastewaters into a source of electricity. The efficient electron transfer between the micro-organism and the anode of the microbial fuel cell plays a major role in the performance of the fuel cell. The organic molecules present in the wastewater posses a certain amount of chemical energy, which is released when converting them to simpler molecules like CO2. The microbial fuel cell is thus a device that converts the chemical energy present in water-soluble waste into electrical energy by the catalytic reaction of microorganisms. Carbon nanotubes (CNTs) have chemical stability, good mechanical properties and high surface area, making them ideal for the design of sensors and provide very high surface area due to its structural network.

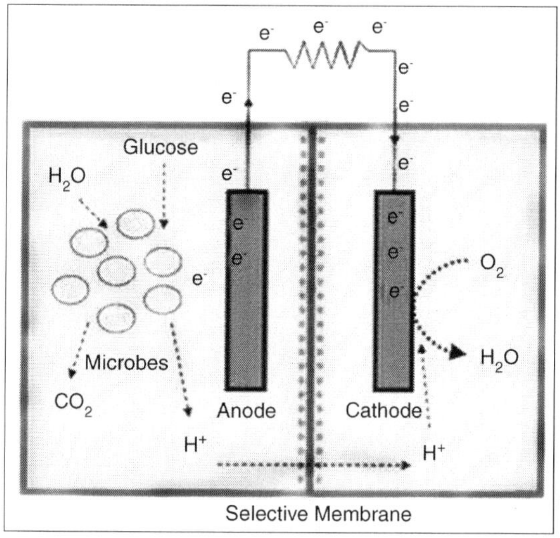

Fig. Schematic Representation of Microbial Fuel Cell

Since carbon nanotubes are also suitable supports for cell growth, electrodes of microbial fuel cells can be built using of CNT. Due to three-dimensional architectures and enlarged electrode surface area for the entry of growth medium, bacteria can grow and proliferate and get immobilized. Multi walled CNT scaffolds could offer self-supported structure with large surface area through which hydrogen producing bacteria (*e.g.*, *E. coli*) can eventually grow and proliferate. Also CNTs and MWCNTs have been reported to be biocompatible for different eukaryotic cells. The efficient proliferation of hydrogen producing bacteria throughout an electron conducting scaffold of CNT can form the basis for the potential application as electrodes in MFCs leading to efficient performance.

SURFACE AREA OF CATALYSIS

Higher surface area available with the nanomaterial counterparts, nano-catalysts tend to have exceptional surface activity. For example, reaction rate at nano-aluminum can go so high, that it is utilized as a solid-fuel in rocket propulsion, whereas the bulk aluminum is widely used in utensils. Nano-aluminum becomes highly reactive and supplies the required thrust to send off pay loads in space. Similarly, catalysts assisting or retarding the reaction rates are dependent on the surface activity, and can very well be utilized in manipulating the rate-controlling step.

PHOSPHORS FOR HIGH-DEFINITION TV

The resolution of a television, or a monitor, depends greatly on the size of the pixel. These pixels are essentially made of materials called "phosphors," which glow when struck by a stream of electrons inside the cathode ray tube (CRT). The resolution improves with a reduction in the size of the pixel, or the phosphors. Nanocrystalline zinc selenide, zinc sulfide, cadmium sulfide, and lead telluride synthesized by the sol-gel techniques are candidates for improving the resolution of monitors. The use of nanophosphors is envisioned to reduce the cost of these displays so as to render highdefinition televisions (HDTVs) and personal computers affordable to be purchase.

NEXT-GENERATION COMPUTER CHIPS

The microelectronics industry has been emphasizing miniaturization, whereby the circuits, such as transistors, resistors, and capacitors, are reduced in size. By achieving a significant reduction in their size, the microprocessors, which contain these components, can run much faster, thereby enabling computations at far greater speeds. However, there are several technological impediments to these advance-ments, including lack of the ultrafine precursors to manufacture these compo-nents; poor dissipation of tremendous amount of heat generated by these microprocessors due to faster speeds; short mean time to failures (poor reliability), etc. Nanomaterials help the industry break these barriers down by providing the manufacturers with nanocrystalline starting materials, ultra-high purity materials, materials with better thermal conductivity, and longer-lasting, durable interconnections (connections between various components in the microprocessors).

Example

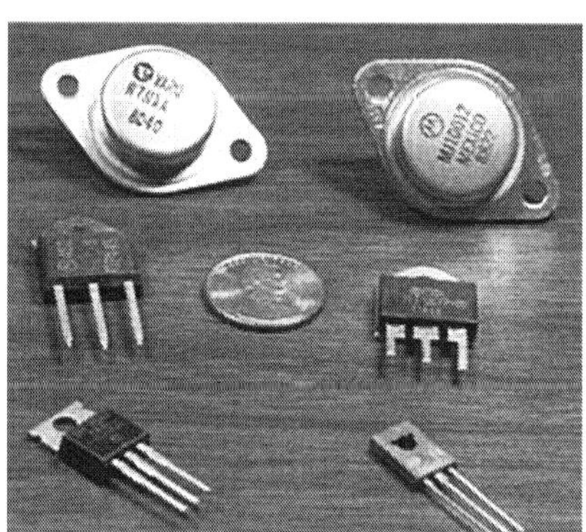

Fig. Silicon Nanowires in Junctionless Transistors

Nanowires for junctionless transistors:

- Transistors are made so tiny to reduce the size of sub assemblies of electronic systems and make smaller and smaller devices, but it is difficult to create high-quality junctions. In particular, it is very difficult to change the doping concentration of a material over distances shorter than about 10 nm. Researchers have succeeded in making the junction-less transistor having nearly ideal electrical properties. It could potentially operate faster and use less power than any conventional transistor on the market today. The device consists of a silicon nanowire in which current flow is perfectly controlled by a silicon gate that is separated from the nanowire by a thin insulating layer. The entire silicon nanowire is heavily n-doped, making it an excellent conductor. However, the gate is p-doped and its presence has the effect of depleting the number of electrons in the region of the nanowire under the gate. The device also has near-ideal electrical properties and behaves like the most perfect of transistors without suffering from current leakage like conventional devices and operates faster and using less energy.

ELIMINATION OF POLLUTANTS

Nanomaterials possess extremely large grain boundaries relative to their grain size. Hence, they are very active in terms of their chemical, physical, and mechanical properties. Due to their enhanced chemical activity, nanomaterials can be used as catalysts to react with such noxious and toxic gases as carbon monoxide and nitrogen oxide in automobile catalytic converters and power generation equipment to prevent environmental pollution arising from burning gasoline and coal.

SUN-SCREEN LOTION

Prolonged UV exposure causes skin-burns and cancer. Sun-screen lotions containing nano-TiO2 provide enhanced sun protection factor (SPF) while eliminating stickiness. The added advantage of nano skin blocks (ZnO and TiO2) arises as they protect the skin by sitting onto it rather than penetrating into the skin. Thus they block UV radiation effectively for prolonged duration. Additionally, they are transparent, thus retain natural skin colour while working better than conventional skin-lotions.

SENSORS

Sensors rely on the highly active surface to initiate a response with minute change in the concentration of the species to be detected. Engineered monolayers (few Angstroms thick) on the sensor surface are exposed to the environment and the peculiar functionality (such as change in potential as the CO/anthrax level is detected) is utilized in sensing.

PROPERTIES OF NANOMATERIALS

Nanomaterials have the structural features in between of those of atoms and the bulk materials. While most microstructured materials have similar properties to the corresponding bulk materials, the properties of materials with nanometer dimensions are significantly different from those of atoms and bulks materials.

This is mainly due to the nanometer size of the materials which render them:

- Large fraction of surface atoms;
- High surface energy;
- Spatial confinement;
- Reduced imperfections, which do not exist in the corresponding bulk materials.

Due to their small dimensions, nanomaterials have extremely large surface area to volume ratio, which makes a large to be the surface or interfacial atoms, resulting in more "surface" dependent material properties. Especially when the sizes of nanomaterials are comparable to length, the entire material will be affected by the surface properties of nanomaterials. This in turn may enhance or modify the properties of the bulk materials. For example, metallic nanoparticles can be used as very active catalysts. Chemical sensors from nanoparticles and nanowires enhanced the sensitivity and sensor selectivity.

The nanometer feature sizes of nanomaterials also have spatial confinement effect on the materials, which bring the quantum effects. The energy band structure and charge carrier density in the materials can be modified quite differently from their bulk and in turn will modify the electronic and optical properties of the materials. For example, lasers and light emitting diodes (LED) from both of the quantum dots and quantum wires are very promising in the future optoelections. High density information storage using quantum dot devices is also a fast developing area. Reduced imperfections are also an important factor in determination of the properties of the nanomaterials.

Nanosturctures and Nanomaterials favours of a selfpurification process in that the impurities and intrinsic material defects will move to near the surface upon thermal annealing. This increased materials perfection affects the properties of nanomaterials. For example, the chemical stability for certain nanomaterials may be enhanced, the mechanical properties of nanomaterials will be better than the bulk materials. The superior mechanical properties of carbon nanotubes are well known. Due to their nanometer size, nanomaterials are already known to have many novel properties. Many novel applications of the nanomaterials rose from these novel properties have also been proposed.

IMPACT OF NANOMATERIALS

Although nanotechnology will help to reduce environmental and human health hazards, for example through better solar cells, replacement of hazardous chemicals, improved pollutant extraction or prevention technologies, unfavourable safety aspects and concerns need also be addressed. However, this is a complex task, as differently shaped nanoparticles of the same substance may have very different properties. Looking beyond the potential technical risks associated with nanomaterials, there is actually only scarce information available about the impact of nanomaterials on non-human species, on ecosystems or the global environment. Even within established questions of toxicology we still do not know how different nanoparticles exactly interact in the human body or the environment, which means that more data is needed.

There is a particular need to address the potential impact of nanomaterials along the lifecycle of a product, from manufacture through disposal ("cradle to grave"). And this 'life cycle view' is not only necessary regarding possible ecotoxicological effects in the different stages of the life cycle of the product, but also regarding the (dissipative) use or consumption of energy and materials (which is in the focus of life cycle assessment (LCA)).

Another problem arises from the fact that the existing relevant regulatory framework (chemicals, wastes, safety) may be adequate for some specific areas, where only small amounts of nanomaterials are used (like in research laboratories), but may not be adequate for the industrial mass production and use of these materials. European legislation1 obliges employers to take measures necessary to ensure the safety and health protection of workers, which should also apply in the case of handling nanomaterials. There are specific Directives relating to the risks to workers due to exposure to carcinogens and mutagens2, chemical agents3, the use of workplace equipment and the use of personal protective equipment5. Under the new REACH chemical regulations, manufacturers have a duty to ensure that substances placed on the market do not adversely affect human health and the environment.

While there is no explicit mention of nanomaterials, they would in principle be covered by these regulations. The Integrated Pollution Prevention and Control Directive requires industrial installations to limit emissions and would be applicable also to nanomaterials. The Waste Directive requires Member States to take measures to ensure that waste treatment does not have an adverse impact on health and the environment.

Consequently, there is a need to evaluate the extent to which existing regulatory and associated risk assessment frameworks (strategies, methodologies and tools), can be applied, or extended to cover mass production, application, and final disposal of nanomaterials and associated products. Looking at REACH there must be a solution regarding the fact, that some important regulations in REACH are triggered by mass (in tons).

This is inadequate for nanomaterials, because for some engineered nanoparticles their number and/or shape could be of higher relevance. Additionally there is a demand for adaptation of the OECD-Guidelines (for hazardous (chemical) substances) to the testing of nanomaterials. But we cannot wait until all these data will be collected. We need processes for 'preliminary assessment' that deliver 'preliminary information as the base of precautionary measures. The latter may be based on information about chemical and or physical properties, on quantitative structure activity relations (QSAR), on the probability of exposure (*e.g.* because of extreme mobility and or persistence), and measures based on it may reach from precautionary measures, aiming at exposure avoidance or reduction, up to principles for a safer (or sustainable) design of nanomaterials and products. Along with the growing (eco)toxicological knowledge about impacts, this 'preliminary knowledge' will become less important.

When working with nanomaterials, there are four main aspects of their life-cycle that must be considered: material selection, manufacturing, application, and disposal/recycle. Although most work focuses on the possible toxic effects of nanocomponents after exposure for risk assessment, it is worth examining the potential contribution of these materials to all impacts listed above when they are added to products or processes, to better understand the importance of the underlying choices that are involved with the implementation of a nanotechnology.

Therefore, understanding the toxicity of nanomaterials and nano-enabled products is important for human and environmental health and safety as well as public acceptance. In numerous toxicological studies it was shown that nanoparticles have implications on the human health inducing, e. g., pulmonary and systemic inflammation. It was also shown that nanoparticles inhaled can translocate to different parts within the human body, including the brain. Data available on the (eco)toxicity of nanomaterials is still limited, but studies prove that there are toxic effects on wildlife and a potential for bioaccumulation in various organisms.

DESTINY OF NANOMATERIALS

Our current knowledge and the available scientific data on nanoparticles characterisation, detection and measurement, their transport, toxicology, exposure and persistence is still insufficient, to allow an accurate and reliable assessment of their final fate in the environment.

But the increasing amount of nanomaterials produced world-wide raises issues on their destiny when released into the environment and on possible hazards due to accumulation in animals, plants and the human body. Metallic nanoparticles may be extremely resistant to degradation and may accumulate in waters or soils. They may aggregate, which in turn will change their properties compared to single nanoparticles, be transported and accumulate in soils, groundwater and sediments. Existing regulations are based on parametres that may not be appropriate for nanoparticles in solution or in suspension.

Concentration data alone is inadequate to quantify the true exposure to nanomaterials, but needs accurate measurement of other nano-specific parametres, like surface area and reactivity. More appropriate analytical methods are required to reliably detect nanoparticles in various compartments and measure their physico-chemical properties in air, water, soil, and consumer products, the media in which humans and ecosystems are exposed to. Better analytical techniques are required to detect these particles in cells, fluids, or plant tissues.

Existing methods to assess environmental exposure levels are not appropriate to determine their environmental fate and current risk assessment procedures need to be modified for adequate hazard characterization. There is a distinct need of data on the properties, toxicokinetics and degradability of nanoparticles to better understand where, in which form and to what extent nanoparticles will end up in the environment, to develop more accurate impact assessment models and to find efficient solutions for product design that favours their reuse and recyclability.

The proposed NanoSustain project will address this need for more reliable data on the fate of nanoparticles by developing technical solutions and advanced computer-based models to support the assessment of the hazard, distribution and fate of nanomaterials that may be released from products.

NANOMATERIALS ESTABLISHMENT

Some nanomaterials occur naturally, but of particular interest are engineered nanomaterials (EN), which are designed for, and already being used in many commercial products and processes. They can be found in such things as sunscreens, cosmetics, sporting goods, stain-resistant clothing, tires, electronics, as well as many other everyday items, and are used in medicine for purposes of diagnosis, imaging and drug delivery. Engineered nanomaterials are resources designed at the molecular (nanometre) level to take advantage of their small size and novel properties which are generally not seen in their conventional, bulk counterparts. The two main reasons why materials at the nano scale can have different properties are increased relative surface area and new quantum effects. Nanomaterials have a much greater surface area to volume ratio than their conventional forms, which can lead to greater chemical reactivity and affect their strength. Also at the nano scale, quantum effects can become much more important in determining the materials properties and characteristics, leading to novel optical, electrical and magnetic behaviours.

Nanomaterials are already in commercial use, with some having been available for several years or decades. The range of commercial products available today is very broad, including stain-resistant and wrinkle-free textiles, cosmetics, sunscreens, electronics, paints and varnishes. Nanocoatings and nanocomposites are finding uses in Book title 1.3 diverse consumer products, such as windows, sports equipment, bicycles and automobiles. There are novel UV-blocking coatings on glass bottles which protect beverages from damage by sunlight, and longer-lasting tennis balls using butylrubber/nano-clay composites. Nanoscale titanium dioxide, for instance, is finding applications in cosmetics, sun-block creams and self-cleaning windows, and nanoscale silica is being used as filler in a range of products, including cosmetics and dental fillings.

ADVANCES IN NANOMATERIALS

The history of nanomaterials began immediately after the big bang when Nanostructures were formed in the early meteorites. Nature later evolved many other Nanostructures like seashells, skeletons etc. Nanoscaled smoke particles were formed during the use of fire by early humans. The scientific story of nanomaterials however began much later. One of the first scientific report is the colloidal gold particles synthesised by Michael Faraday as early as 1857. Nanostructured catalysts have also been investigated for over 70 years. By the early 1940's, precipitated and fumed silica nanoparticles were being manufactured and sold in USA and Germany as substitutes for ultrafine carbon black for rubber reinforcements. Nanosized amorphous silica particles have found large-scale applications in many everyday consumer products, ranging from non-diary coffee creamer to automobile tires, optical fibres and catalyst supports. In the 1960s and 1970's metallic nanopowders for magnetic recording tapes were developed. In 1976, for the first time, nanocrystals produced by the now popular inert-gas evaporation technique was published by Granqvist and Buhrman. Recently it has been found that the Maya blue paint is a nanostructured hybrid material. The origin of its colour and its resistance to acids and biocorrosion are still not understood but studies of authentic samples from Jaina Island show that the material is made of needle-shaped palygorskite (clay) crystals that form a superlattice with a period of 1.4 nm, with intercalates of amorphous silicate substrate containing inclusions of metal (Mg) nanoparticles. The beautiful tone of the blue colour is obtained only when both these nanoparticles and the superlattice are present, as has been shown by the fabrication of synthetic samples.

NANOMATERIALS ENGINEERING

Today nanophase engineering expands in a rapidly growing number of structural and functional materials, both inorganic and organic, allowing to manipulate mechanical, catalytic, electric, magnetic, optical and electronic functions.

The production of nanophase or cluster-assembled materials is usually based upon the creation of separated small clusters which then are fused into a bulk-like material or on their embedding into compact liquid or solid matrix materials. *e.g.* nanophase silicon, which differs from normal silicon in physical and electronic properties, could be applied to macroscopic semiconductor processes to create new devices.

For instance, when ordinary glass is doped with quantised semiconductor "colloids," it becomes a high performance optical medium with potential applications in optical computing.

CLASSIFICATION OF NANOMATERIALS

Nanomaterials have extremely small size which having at least one dimension 100 nm or less. Nanomaterials can be nanoscale in one dimension (*e.g.* surface films), two dimensions (*e.g.* strands or fibres), or three dimensions (*e.g.* particles). They can exist in single, fused, aggregated or agglomerated forms with spherical, tubular, and irregular shapes. Common types of nanomaterials include nanotubes, dendrimers, quantum dots and fullerenes.

Nanomaterials have applications in the field of nano technology, and displays different physical chemical characteristics from normal chemicals (*i.e.*, silver nano, carbon nanotube, fullerene, photocatalyst, carbon nano, silica).

Classification

Materials referred to as "nanomaterials" generally fall into two categories: fullerenes, and inorganic nanoparticles.

Fullerenes

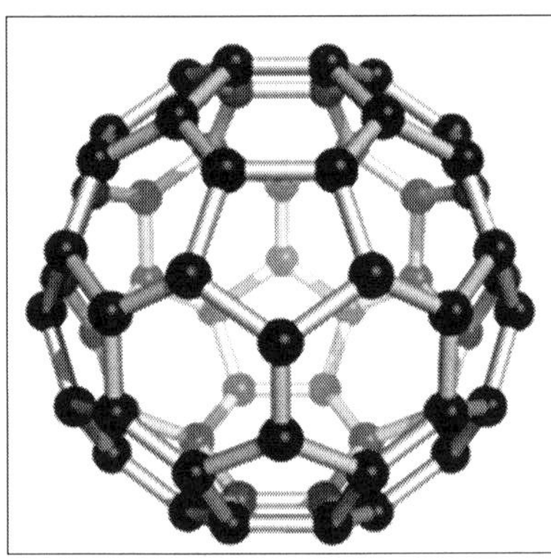

Fig. Rotating View of Buckminsterfullerene C_{60}

The fullerenes are a class of allotropes of carbon which conceptually are graphene sheets rolled into tubes or spheres. These include the carbon nanotubes (or silicon nanotubes) which are of interest both because of their mechanical strength and also because of their electrical properties.

For the past decade, the chemical and physical properties of fullerenes have been a hot topic in the field of research and development, and are likely to continue to be for a long time. In April 2003, fullerenes were under study for potential medicinal use: binding specific antibiotics to the structure of resistant bacteria and even target certain types of cancer cells such as melanoma. The October 2005 issue of Chemistry and Biology contains an article describing the use of fullerenes as light-activated antimicrobial agents. In the field of nanotechnology, heat resistance and superconductivity are among the properties attracting intense research. A common method used to produce fullerenes is to send a large current between two nearby graphite electrodes in an inert atmosphere. The resulting carbon plasma arc between the electrodes cools into sooty residue from which many fullerenes can be isolated. There are many calculations that have been done using ab-initio Quantum Methods applied to fullerenes. By DFT and TDDFT methods one can obtain IR, Raman and UV spectra. Results of such calculations can be compared with experimental results.

Sol-gel

The sol-gel process is a wet-chemical technique commonly used to synthesise a wide variety of nanomaterials.

CHARACTERIZATION

The first observations and size measurements of nanoparticles were made during the first décade of the 20th century. They are mostly associated with the name of Zsigmondy who made detailed studies of gold sols and other nanomaterials with sizes down to 10 nm and less. He published a book in 1914. He used an ultramicroscope that employs a *dark field* method for seeing particles with sizes much less than light wavelength.

There are traditional techniques developed during 20th century in Interface and Colloid Science for characterizing nanomaterials. These are widely used for *first generation* passive nanomaterials.

These methods include several different techniques for characterizing particle size distribution. This characterization is imperative because many materials that are expected to be nano-sized are actually aggregated in solutions. Some of methods are based on light scattering. Other apply ultrasound, such as ultrasound attenuation spectroscopy for testing concentrated nano-dispersions and microemulsions. There is also a group of traditional techniques for characterizing surface charge or zeta potential of nanoparticles in solutions. This information is required for proper system stabilzation, preventing its aggregation or flocculation. These methods include microelectrophoresis, electrophoretic light scattering and electroacoustics. The last one, for instance colloid vibration current method is suitable for characterizing concentrated systems.

NANOMATERIALS FOR THE FUTURE

A conceptual high-energy-density physics target about 2 millimetres in diameter for use at Livermore's National Ignition Facility. Deuterium gas is encapsulated in high-atomic-number doped foam and deuterium ice and encased in an outer beryllium–copper shell. Less than 20 years ago, nanomaterials—

substances on the scale of one-billionth of a meter—were a curiosity. Today, their diverse technological applications seem to be endless. Materials reduced to the nanoscale show very different physical, chemical, and mechanical properties compared to those they exhibit on a macroscale. For example, nanoparticles have electronic structures that lie between that of bulk materials and atomic or molecular structures. Nanomaterials have real potential for impact in applications such as alternative energy conversion and energy harvesting.

The development of new nanomaterials will also lead to new mass-produced consumer products in cosmetics, cars, building components, clothes, health, and new smart electronic devices. Novel assembly and manipulation of nanostructured materials in confined geometries with tailored composition and function will have applications in catalysis, sensing, hydrogen storage, advanced nuclear materials, corrosion-resistant coatings, and photonics. The design of new nanomaterials with novel properties and functionalities requires atomic-level control of the material's structure. The fabrication of functional nanomaterials can be achieved by conventional top-down methods such as lithography or bottom-up approaches such as self-assembly. The latter approach can be very cost-effective for the fabrication of macroscopic three-dimensional nanostructures.

However, self-assembly is scientifically very challenging because it requires detailed understanding of materials interactions at the atomic level. The goal of this LDRD project is twofold. The research team is advancing the scientific field of functional nanomaterials by combining the best of directed self-assembly with automated lithography and assembly to develop nanomaterials for fusion-energy applications for advanced laser systems. Second, they are applying this scientific knowledge to fabricate complex laser-fusion target structures by high-throughput techniques such as automated assembly and templating. Success will lay the scientific foundation for efficient production of laser fusion targets with unprecedented complexity.

A successful experiment at fusion-class laser facilities such as the National Ignition Facility creates a unique laboratory environment previously only available in a nuclear test. First among the unique characteristics is an inertially confined burning hydrogen plasma. Second is the bright neutron source that such a plasma will create. The challenge is to create laser targets that take advantage of this environment. To carry out experiments in this environment, researchers must place nanostructured materials with controlled composition and precise geometry in cubic millimetre spaces inside an inertial-confinement fusion capsule. Inertial fusion ignition will enable important laboratory and national missions, as well as new science. A series of such experiments is proposed for the National Ignition Facility, and NNSA's top scientific priority is significant advancement in the understanding of boost physics within the next five years. Thus, among the first proposed experiments is a suite of boost physics experiments. These are designed to investigate the mix in burning plasmas. Complex targets for study of the mix in burning hydrogen plasmas, nuclear physics in high-neutron-brightness environments, and fast-ignition inertial-confinement fusion require precisely placing nanoporous materials with small amounts of additional incorporated elements to alter the material's properties (a process known as doping), inside the target capsule.

Traditionally, target capsules are made by assembling prefabricated and machined components of the structure. This approach limits the complexity of the structures that can be achieved. With self-assembly techniques, complex targets could be built either from the inside out or outside in. In the latter case, the ablator shell with a fill hole would serve as both a template and a container in which materials could be grown or modified. LDRD researchers will use this "chemistry in a capsule" approach to generate complex nanostructures within the ablator shell by in situ self-assembly of functional building blocks.

For example, a uniform and conforming film of a porous material lining the inside of the shell could be prepared by spin coating and subsequent curing by light, heat, or chemical initiation. Once formed, the properties of a nanoporous capsule liner could be further fine-tuned by atomic layer deposition. Possible applications include doping with mid-to-high-atomic-number elements as well as surface engineering of the wetting behaviour. In principle, this self-assembly technique could be applied to any nanostructure formation.

In just the first year of this project, the LDRD team has made significant progress in all these areas. They have successfully demonstrated the feasibility of the chemistry in a capsule approach by loading a diamond laser target with the nanomaterial precursor solution and forming nanoporous foam inside the capsule. Team members installed an atomic-layer-deposition system and demonstrated that the technique can be applied to deposit mid-to-high-atomic-number elements on the nanoporous foam liner inside the capsule. They also demonstrated that the capsule liner can be rendered inactive and then reactivated for atomic layer deposition.

This discovery opens the door to generating nanoporous foams with three-dimensional doping patterns by area-selective atomic layer deposition. The effects and requirements of this tailored doping on yield and symmetry of foam-filled capsules was assessed by hydrodynamic simulations. X-ray scattering was used to study the effects of wetting organic aerogels with liquid hydrogen on the foam morphology. These experiments demonstrated that even low-density aerogels can survive liquid hydrogen wetting and de-wetting cycles. Finally, an in situ film-stress-monitoring capability was developed to study the microstructure-stress correlation in thick metal films grown by magnetron sputtering. The team was able to control the texture—and thus the stress of thick beryllium films—by applying a bias to the substrate, thereby changing the initial stress state from tensile to compressive. Besides this impressive progress, they also developed novel nanomaterials for energy storage and conversion applications and studied their unique properties.

Future LDRD efforts will include casting of thin films of aerogels in target capsules by growing cross-linkable polymer brushes, for example. Using in situ x-ray techniques, the team will study the effect of aerogel morphology on wetting with liquid hydrogen.

They will develop novel atomic-layer-deposition precursor chemistries for doping with high-atomic-number materials. Area-selective doping will be attempted through spatially selective modifications of the substrate's surface chemistry. In addition, new functionalities will be added to their nanomanipulator to measure thickness profiles from three-dimensional targets and to apply picoliter amounts of liquids. Researchers will continue work on stress control in metal deposition and develop predictive multiscale tools.

OPTICAL PROPERTIES

One of the most fascinating and useful aspects of nanomaterials is their optical properties. Applications based on optical properties of nanomaterials include optical detector, laser, sensor, imaging, phosphor, display, solar cell, photocatalysis, photoelectrochemistry and biomedicine. The optical properties of nanomaterials depend on parameters such as feature size, shape, surface characteristics, and other variables including doping and interaction with the surrounding environment or other nanostructures. Likewise, shape can have dramatic influence on optical properties of metal nanostructures.

Figure exemplifies the difference in the optical properties of metal and semiconductor nanoparticles. With the CdSe semiconductor nanoparticles, a simple change in size alters the optical properties of the nanoparticles. When metal nanoparticles are enlarged, their optical properties change only slightly as observed for the different samples of gold nanospheres in figure. However, when an anisotropy is added to the nanoparticle, such as growth of nanorods, the optical properties of the nanoparticles change dramatically.

ELECTRICAL PROPERTIES

"Electrical Properties of Nanoparticles" discuss about fundamentals of electrical conductivity in nanotubes and nanorods, carbon nanotubes, photoconductivity of nanorods, electrical conductivity of nanocomposites. One interesting method which can be used to demonstrate the steps in conductance is the mechanical thinning of a nanowire and measurement of the electrical current at a constant applied voltage. The important point here is that, with decreasing diameter of the wire, the number of electron wave modes contributing to the electrical conductivity is becoming increasingly smaller by well-defined quantised steps. In electrically conducting carbon nanotubes, only one electron wave mode is observed which transport the electrical current.

As the lengths and orientations of the carbon nanotubes are different, they touch the surface of the mercury at different times, which provides two sets of information:

- The influence of carbon nanotube length on the resistance; and
- The resistances of the different nanotubes.

As the nanotubes have different lengths, then with increasing protrusion of the fibre bundle an increasing number of carbon nanotubes will touch the surface of the mercury droplet and contribute to the electrical current transport.

MECHANICAL PROPERTIES

"Mechanical Properties of Nanoparticles" deals with bulk metallic and ceramic materials, influence of porosity, influence of grain size, superplasticity, filled polymer composites, particle-filled polymers, polymer-based nanocomposites filled with platelets, carbon nanotube-based composites. The discussion of mechanical properties of nanomaterials is, in to some extent, only of quite basic interest, the reason being that it is problematic to produce macroscopic bodies with a high density and a grain size in the range of less than 100 nm. However, two materials, neither of which is produced by pressing and sintering, have attracted much greater interest as they will undoubtedly achieve industrial importance.

These materials are polymers which contain nanoparticles or nanotubes to improve their mechanical behaviours, and severely plastic-deformed metals, which exhibit astonishing properties. However, because of their larger grain size, the latter are generally not accepted as nanomaterials. Experimental studies on the mechanical properties of bulk nanomaterials are generally impaired by major experimental problems in producing specimens with exactly defined grain sizes and porosities. Therefore, model calculations and molecular dynamic studies are of major importance for an understanding of the mechanical properties of these materials. Filling polymers with nanoparticles or nanorods and nanotubes, respectively, leads to significant improvements in their mechanical properties. Such improvements depend heavily on the type of the filler and the way in which the filling is conducted.

The latter point is of special importance, as any specific advantages of a nanoparticulate filler may be lost if the filler forms aggregates, thereby mimicking the large particles. Particulatefilled polymer-based nanocomposites exhibit a broad range of failure strengths and strains. This depends on the shape of the filler, particles or platelets, and on the degree of agglomeration. In this class of material, polymers filled with silicate platelets exhibit the best mechanical properties and are of the greatest economic relevance. The larger the particles of the filler or agglomerates, the poorer are the properties obtained.

Although, potentially, the best composites are those filled with nanofibres or nanotubes, experience teaches that sometimes such composites have the least ductility. On the other hand, by using carbon nanotubes it is possible to produce composite fibres with extremely high strength and strain at rupture. Among the most exciting nanocomposites are the polymerceramic nanocomposites, where the ceramic phase is platelet-shaped. This type of composite is preferred in nature, and is found in the structure of bones, where it consists of crystallised mineral platelets of a few nanometers thickness that are bound together with collagen as the matrix. Composites consisting of a polymer matrix and defoliated phyllosilicates exhibit excellent mechanical and thermal properties.

MAGNETIC PROPERTIES

Bulk gold and Pt are non-magnetic, but at the nano size they are magnetic. Surface atoms are not only different to bulk atoms, but they can also be modified by interaction with other chemical species, that is, by capping the nanoparticles. This phenomenon opens the possibility to modify the physical properties of the nanoparticles by capping them with appropriate molecules. Actually, it should be possible that non-ferromagnetic bulk materials exhibit ferromagnetic-like behaviour when prepared in nano range. One can obtain magnetic nanoparticles of Pd, Pt and the surprising case of Au (that is diamagnetic in bulk) from non-magnetic bulk materials. In the case of Pt and Pd, the ferromagnetism arises from the structural changes associated with size effects.

However, gold nanoparticles become ferromagnetic when they are capped with appropriate molecules: the charge localised at the particle surface gives rise to ferromagnetic-like behaviour. Surface and the core of Au nanoparticles with 2 nm in diameter show ferromagnetic and paramagnetic character, respectively. The large spin-orbit coupling of these noble metals can yield to a large anisotropy and therefore exhibit high ordering temperatures. More surprisingly, permanent magnetism was observed up to room temperature for thiol-capped Au nanoparticles. For nanoparticles with sizes below 2 nm the localised carriers are in the 5d band. Bulk Au has an extremely low density of states and becomes diamagnetic, as is also the case for bare Au nanoparticles. This observation suggested that modification of the d band structure by chemical bonding can induce ferromagnetic like character in metallic clusters.

4

Chemicals in the Environment

CHEMICALS AND THEIR CHEMISTRY

Everything around us - animal, vegetable and mineral - is made up of chemicals. Some are simple substances such as the water molecule, which is made up of hydrogen and oxygen atoms, and others are very complicated compounds made up of many different chemical elements.

As well as naturally occurring chemical compounds, scientists have created millions more in the past 100 years that do not exist in nature. Chemicals come as powders, pellets, dusts, liquids, vapours, gases, etc.

Chemists tend to divide the world into inorganic and organic compounds, largely depending on whether they contain the element carbon.

Carbon-containing compounds were originally described as organic because many of them came from nature. Since then, chemists have synthesised many organic compounds, many of which are important in industrial products and processes. The organic solvents and organophosphate pesticides involved in the case studies are just a few examples of these organic compounds. Inorganic materials, comprising metals, minerals and salts, among other types, are much less likely to contain carbon.

WHAT'S IN A NAME?

Individual chemical compounds are described by their formulae, such as H_2O for water and CH_4 for methane. A chemical formula describes the number of atoms of each element making up the compound, whereas the structure describes the arrangement of these atoms. The molecule is not flat but a three-dimensional tetrahedron, with the carbon atom in the centre and a hydrogen at each of the four corners.

Chemical structure is not just of academic interest. It affects how chemicals react with each other, and with biological systems, including humans and the environment. It is common for chemicals with the same formula to have different structures (called isomers), and therefore have different properties and toxicity. In fact, the seemingly small difference of one molecule being the mirror image of the other (called stereoisomers) can cause major differences in their biological effects. Although certain rules apply to naming chemicals, most chemicals are known by more than one name, or synonym, and some also have trade names. This means that finding information on specific chemicals can appear complicated but useful information can be gleaned by non-chemists and this becomes easier with familiarity and practice.

When searching for information on a chemical, it is always useful to know its CAS registry number. This number is assigned by the US Chemical Abstracts Service, and each chemical has its own unique number. This system is accepted globally. Many books and databases of chemicals are indexed by CAS registry number.

PROPERTIES

As well as their formula and structure, chemists describe a great many properties of a chemical. Some of these are also important to the toxicologist or industrial hygienist because they tell us how a chemical behaves in various situations.

According to its boiling point and melting point, a chemical will be a solid, liquid or gas at room temperature and pressure. This is known as the chemical's physical state, and this changes with temperature and pressure. Water, for example, changes from a solid to a liquid to a gas between 0 and 100°C. A chemical's physical state, coupled with other properties such as its volatility, will influence the likelihood of its entry into the human body by various routes of exposure.

Several other properties influence the effect of chemicals:

- Vapour density: The weight of a volume of a gaseous chemical compared with the weight of the same volume of air. If the vapour density is greater than 1, the gas is heavier than air and will tend to collect at floor level where it can be a fire or explosion hazard. In confined spaces, gases with vapour density less than 1 may replace air in the space, so that there is a danger of suffocation.
- Vapour pressure: Describes how fast a solid or liquid evaporates, and increases with increasing temperature. A liquid which evaporates easily into the air is more likely to be inhaled.
- Flash point: Describes the temperature at which a substance gives off enough vapour to form a mixture with air which can be ignited by a spark or flame.
- *Autoignition temperature*: The lowest temperature at which a substance burns without a spark or flame.
- *Explosive or flammability V limits*: An upper and lower concentration of a gas or vapour in air, between which it may explode if ignited by a spark or flame.

TOXICITY

How poisonous - or toxic - a chemical is depends on the nature of the chemical itself, including its structure and properties. But its potential to cause harm to workers also depends on the dose and duration of exposure, as well as other factors. The word 'toxic' comes from the Greek toxicon or arrow-poison. The word toxic is the root of many terms covered in the following sections, such as toxicity (how toxic a substance is) and toxicology (the study of poisons).

"Toxicology is the study of the harmful effects of chemicals on biological systems. It is a hybrid science built on advances in biochemistry, physiology, pathology, physical chemistry, pharmacology, and public health." "It is probably safe to say that the mechanism of action of no chemical is understood in every detail. Toxicologists know a great deal about a few chemicals, a little about many, and next to nothing about most".

The following sections describe how chemicals enter the body (routes of exposure), and the effects that they can have on various parts of the body. We will look at the time-scale over which these effects operate, so-called acute and chronic effects, and how certain combinations of chemicals are much more toxic together than would be predicted by adding up the damage they do on their own. Finally, we will look at the many ways in which scientists study and measure toxicity, so that toxicity information in books, journals and safety data sheets can be interpreted more easily.

Routes of Exposure

Chemicals enter the body by three main routes: through the lungs (inhalation); through the skin (dermal absorption); and by being swallowed (ingestion). They can also enter through the eyes. Chemicals can have local effects, such as damaging the skin, but may also be absorbed into the body and affect other, distant parts of the body (systemic effects).

The Lungs

The job of the lungs is to exchange gases, allowing oxygen to be absorbed into the blood stream from the air we breathe, and get rid of waste carbon dioxide from the body. In the same way, the lungs will readily absorb other chemicals found in workplace air. As well as gases, solids can reach the lungs when present as small particles suspended in air, such as dusts and fumes. One of our case studies involves fumes from solder flux, and dusts from asbestos, silica, wood, cotton and flour have killed thousands of workers. Some dusts may also cause a particular kind of cxplosion.

The Skin

Although the skin acts as a protective barrier against many micro-organisms and chemicals, some chemicals can penetrate the skin and enter the blood stream. Whether or not a chemical is absorbed through the skin depends on its structure: chemicals need to be able to dissolve in both water and fat (lipids) to get through the skin. Those that are insoluble, or dissolve only in fats or water, and chemicals made up of very large molecules, tend not to penetrate the skin. Chemicals are more easily absorbed where the skin is thin, such as on the forearms, than through the thick skin covering the palms of the hands and soles of the feet.

Chemicals are also more easily absorbed if skin is moist or damaged. Some chemicals, *e.g.* organic solvents, cause 'defatting' of the skin, making it a less effective barrier against further chemical exposure. Methyl ethyl ketone (MEK), the organic solvent used by Tony Bradshaw, is rapidly absorbed through the skin. In a study using human volunteers, MEK was applied to the skin of their forearms, and could be detected in the air they breathed out just three minutes later. During that time, the MEK had made its way through the skin, into the blood, and then to the lungs. Because MEK is soluble in water, it is absorbed through the skin even faster if the skin is sweaty. However, the amount of MEK in exhaled air accounted for only 10 per cent of the amount applied to their skin. The other 90 per cent was excreted in the urine, both as MEK and as its metabolites.

Ingestion

As well as the lungs and skin, chemicals can also enter the body if swallowed. In the workplace, this can occur if areas used to eat, drink or smoke are contaminated with chemicals, or if workers do not wash their hands or remove their gloves before eating or smoking.

Transportation

Once absorbed into the blood, the chemical is carried around the body in the blood stream, and where it ends up is influenced by its structure and properties. However, some barriers exist in the body which can keep out some (but not other) chemicals, such as the blood-brain barrier which helps protect the brain, and the placenta which helps protect a foetus.

Storage

Inside the body, some chemicals are stored in certain tissues, such as fat or bone, and while they remain bound up there, they may do little damage. However, under certain conditions such as rapid weight loss, large amounts of the chemical are released into the blood. How long such chemicals remain in the body varies, but some, like the pesticide DDT, remain for years. One of the reasons why DDT stopped being used in the developed world was because of this persistence in the environment, and even though it has not been used for years in the developed world, most of us have DDT in our bodies.

Metabolism

If chemicals are not stored, the body deals with them by metabolising (changing their structure) and excreting them. This occurs mainly in the liver, but also the skin, lungs, gut and kidneys, by similar processes used by our bodies to metabolise the chemicals which make up our food.

The products of metabolism are known as metabolites, and these can be more or less toxic than the original chemical. In fact, many of the adverse effects of chemical exposure are due to the effects of metabolites.

The pathways involved in metabolising chemicals vary greatly between species, and also between individuals, which explains why some people are harmed by very low levels of chemicals that others seem able to tolerate.

Excretion and Biological Monitoring

Chemicals and their metabolites are excreted from the body, mainly via urine produced by the kidneys. Small amounts are also excreted by the lungs, and in sweat, semen, milk, saliva and bile. The amount of a chemical a worker has been exposed to can sometimes be estimated by measuring how much of the chemical, or certain metabolites, is found in urine. This is known as biological monitoring.

Toxic Effects

We need to look more at what happens when chemicals enter the body and at the range of effects they might have. Toxicologists often quote a 16th Century Swiss doctor, Paracelsus, to illustrate that it is incorrect to divide chemicals into those which are toxic, and those which are not.

Paracelsus said, "All substances are poisons; there is none which is not a poison. The right dose differentiates a poison and a remedy." Certain substances, including some naturally occurring plant and animal poisons, are lethal at tiny doses. Others, such as some of the things we eat, drink, or work with, can be lethal but only in massive quantities.

Acute and Chronic Exposure

As well as the dose of a chemical, its toxicity also depends on how long exposure lasts, the duration of exposure. This duration is one way that the toxicity of a chemical can be categorised. Single exposures are referred to as acute exposure, and repeated exposure over a longer time as chronic exposure. Duration of exposure should be reported with toxicity data in safety data sheets, etc. The other way toxicity is categorised is according to the chemical's target, the organ or body system which it damages.

The terms 'acute' and 'chronic' are also used to describe how long it takes for the effect of a chemical to occur, and it is important to be aware of these two uses. An acute effect is one which happens immediately on exposure, whereas a chronic effect does not. It may take years to appear. The time between exposure and the onset of disease is referred to as the latency period. Because there is such a long latency between certain chemical exposures and diseases such as cancer, it is often difficult to link them to occupational exposures. This is one reason why workers should insist that the chemicals they work with are recorded on their medical records.

These two uses of the terms 'acute' and 'chronic' are not connected in the sense that an acute exposure only leads to an acute effect. In fact, an acute exposure can lead to either an acute or a chronic effect.

When talking about toxicity, the easiest way to divide chemicals is by the organ or system they damage. These target organs or systems are often referred to in safety data sheets and toxicology books. The commonest are: the lungs, the skin, the gut, the liver, the kidneys, the nervous system, the blood, the cardiovascular system, the immune system, and the reproductive system. There are even chemicals which can affect hearing.

Chemicals causing liver damage are sometimes called 'hepatotoxins', those which damage the kidneys 'renal toxins', and those harming the nervous system 'neurotoxins'. Chemicals that cause cancer, although they may affect either one or several organs, are lumped together and described as 'carcinogens'. Those that cause birth defects are called 'teratogens'.

"A toxin is generally understood to be a substance that is harmful to biological systems, but within this simple concept lies a great deal of variability. A substance that is harmful at a high dose may be innocuous or even essential at a lower dose. A toxin may damage a specific body system, or it may exert a general effect on an organism. A substance that is toxic to one species may not be toxic to another because of different metabolic pathways or protective mechanisms. And the biologic damage may be temporary, permanent over the organism's lifetime, or expressed over subsequent generations"

Respiratory System

The respiratory system includes the nose and the tubes leading to the lungs (the trachea and bronchi), as well as the lungs themselves. Because their job is to exchange gases, the lungs are vulnerable to chemicals both in the form of small particles, and as gases, vapours and mists. The lungs have a surface area of about 500 square feet (Rodricks) and come into contact with 20 kg of air each day (Stacey). Chemicals can cause a variety of damage to the lungs, including irritation, structural changes such as fibrosis, and cancer.

Irritation

Many chemicals irritate the lungs, such as ammonia, chlorine, hydrogen chloride and sulphur dioxide gases. The respiratory system responds to these irritants by tightening the bronchi, which results in a feeling of tightchestedness. This is sometimes referred to as 'dyspnoea' in safety data sheets. If the irritation is very severe, cells can be damaged and fluid released (oedema). Such damage can make the lungs more vulnerable to infection.

Emphysema and Fibrosis

Chronic exposure to some chemicals, such as certain forms of some metals, can cause structural changes in the lungs like emphysema. Dusts can also produce a particular type of lung damage called fibrosis. Some dusts, like crystalline silica, cause cells in the lungs to produce fibrous materials which can build up, making the lung rigid and unable to work properly. Asbestos also causes a kind of fibrosis, known as asbestosis, as well as mesothelioma and lung cancer. As well as asbestos, other chemicals that cause lung cancer include: radon, arsenic, some forms of chromium, nickel, cadmium, bischloromethyl ether (BCME), beryllium, soot and environmental tobacco smoke.

Asphyxia

Some gases, like methane and nitrogen, although they have little effect on the body itself, can displace oxygen in air and cause suffocation or asphyxia. Other chemicals like carbon monoxide work like this but on a cellular level, where they stop cells taking up or using oxygen in the blood.

Asthma

Other chemicals, like colophony, cause allergic reactions in the respiratory system, including occupational asthma. Occupational asthma is the most commonly reported occupational disease according to SWORD, the UK respiratory disease monitoring scheme.

In 1995, a National Asthma Campaign telephone survey of over 300 small and medium-sized businesses found 59 per cent did not know what occupational asthma was, 62 per cent of firms who understood the condition had done nothing to control the hazards, and 81 per cent had taken no action to control exposure to occupational asthmagens.

In a study of 100 patients of Birmingham Heartlands Hospital's occupational lung disease unit, 91 per cent said they had never been informed about the risks of getting asthma at work, and 73 per cent had never seen a safety data sheet. Most worked in the car industry, hospitals, foundries or with wood.

Many worked with chemicals well known to cause asthma, such as isocyanates, colophony and wood dust. Although almost half worked for firms which had an occupational health service, less than one-third had pre-employment screening. The results of this study paint a very different picture of COSHH compliance compared with the HSE's 1991-92 evaluation survey which found 80 per cent of employers provided workers with adequate information and training.

According to the Labour Force Survey, 70,000 people say that their work either caused their asthma, or made it worse. Over 1,000 new cases of occupational asthma are reported to the UK's Surveillance of Work-related and Occupational Respiratory Diseases (SWORD) scheme each year. SWORD was set up in 1989, funded by the HSE, and pools data from chest clinics and occupational physicians.

This is likely to be a significant underestimate and TUC figures suggest that up to 400,000 people suffer from asthma because of their work. Occupational asthma causes over a million days' sick leave each year, four times as many in 1994 as were lost in strikes.

Skin

The skin is the largest organ of the body, covering about 9 m^2, the size of a tennis court. Skin effects are much more obvious than damage to other organs, and in the USA, for example, skin diseases account for a third of all reported occupational diseases (Levy and Wegman).

Occupational skin diseases can be classified into several types, or reaction patterns: contact dermatitis (irritant contact dermatitis and allergic contact dermatitis), urticaria, follicle abnormalities, infections, pigment disorders, and neoplasms.

Dermatitis

Contact dermatitis is the most common occupational skin reaction. In the UK, it accounts for 80 per cent of the cases reported to EPI-DERM, the skin disease monitoring scheme. Dermatitis is an inflammation reaction, where the area which comes into contact with the chemical becomes red, swollen or blistered, and feels itchy or burning.

It can be caused by irritants or result from an allergic reaction. The HSE says: "At worst, contact dermatitis can be as disabling at work as the loss of a limb." The HSE estimates 132,000 working days are lost every year because of occupational dermatitis, and many people are forced to change jobs as a result.

Irritant contact dermatitis can be caused by many substances. Some, like soaps, detergents and many solvents, are mild irritants to which the skin needs to be exposed in large amounts or for long periods to cause dermatitis. Others are very strong irritants, such as hydrogen fluoride and sulphuric acid (both strong acids), or sodium hydroxide (a strong alkali).

These chemicals cause what are sometimes called 'chemical burns', destroying the skin and causing ulcers and scarring. Irritant contact dermatitis is much more successfully treated if detected early.

Allergic contact dermatitis results from a specific allergy, where an individual is sensitised by an initial exposure, and then reacts when re-exposed to even small amounts. Chemicals which often cause allergic contact dermatitis are: latex (in surgical gloves), formaldehyde, rubber additives, nickel or chromium compounds, bactericides and fungicides, adhesives and sealants, plants and wood, and hairdressing chemicals. Because people vary in their susceptibility to substances that can cause allergic contact dermatitis, it may occur in only a few people in a workplace. Just because only one or two people are affected does not mean it is not work-related.

Chloracne

Some chemicals like creosote, oils and greases can make existing acne worse, especially affecting skin beneath clothing which is saturated with the oils. Other chemicals cause a specific type of acne called 'chloracne'. Chloracne is small straw-coloured cysts and inflamed follicles, often behind the ears and at the outer corner of the eyes.

It is caused by chlorinated hydrocarbons, found in herbicide manufacturing and cable splicing, and by polychlorinated biphenyls. As well as its effect on the skin, chloracne should be treated as evidence that these chemicals have been absorbed into the body through the skin.

Pigment Changes

Although changes in skin pigmentation can be due to any skin injury, particular kinds of pigment loss can be caused by chemical exposure. One such is the monobenzyl ether of hydroquinone, an antioxidant used in rubber manufacture.

Skin Cancer

The final, and most potentially serious, kinds of skin disease are neo-plasms (growths). These may be benign or cancerous. Discovery of skin cancer of the scrotums of chimney sweeps in London in 1775 by Sir Percival Potts was the first recorded case of occupational cancer.

Nervous System

The nervous system is usually described in two parts, the central nervous system or CNS (brain and spinal cord), and the peripheral nervous system (the other nerves which control the muscles and senses). Chemicals may affect the CNS or the peripheral nervous system or both.

The Peripheral Nervous System

Chemicals that damage the peripheral nervous system do so in one or both of two ways. Chemicals can damage the outer covering of the nerve (the myelin sheath), which can re-grow quite quickly, or the nerve (axon) itself. Axons can regenerate, but only slowly. The solvents n-hexane and methyl butyl ketone (or more precisely, their metabolite hexane-2,5-dione) are examples of chemicals which damages axons. n-Hexane has been used in paints, glues, varnishes, plastics and rubber and its effects on the peripheral nervous system have been reported in shoe makers and cabinet makers (Stacey).

Damage to the peripheral nervous system causes symptoms like pins and needles (parasthesia), numbness and weakness in the hands and feet. The onset of these symptoms can be delayed for several months. Although many chemicals which affect the peripheral nervous system cause very similar effects, some produce more specific symptoms. Arsenic, for example, causes painful limbs and sensitive feet. Trichloroethylene, an organic solvent, affects nerves in the face, causing facial numbness and weakness, while the insecticide chlordecone (also known as Kepone) causes abnormal eye movements. Recovery from peripheral nervous system damage depends on the chemical involved, and the intensity of the exposure (Levy and Wegman).

The Liver

As well as storing vitamins and iron, regulating blood sugar levels and its role in digestion, the liver metabolises foreign chemicals. Metabolism sometimes results in a chemical being changed into a more toxic metabolite, and this is one way in which chemicals can damage the liver.

As with other organs, liver toxins can be grouped together according to the kind of liver disease they cause, including acute hepatitis (inflammation of the liver) or chronic diseases like cirrhosis and cancer. These diseases can also be caused by viruses such as hepatitis B, an important risk for health care workers, as well as non-occupational factors.

Chemicals that cause acute hepatitis include carbon tetrachloride, chloroform, dinitrophenol, dinitrobenzene, dioxin, polychlorobi-phenyls, the pesticide DDT, chlordecone, chlorobenzenes, the anaesthetic halothane, the dye feedstock methylenedianiline and the explosive TNT. Symptoms of acute hepatitis include headache, nausea, vomiting, dizziness and drowsiness.

As well as alcohol, cirrhosis is also caused by arsenic and vinyl chloride. Vinyl chloride is well known for causing liver angiosarcoma, a rare kind of cancer. After clusters of cases were reported in the 1970s, studies were done in vinyl chloride plants which found high levels of this liver cancer. Vinyl chloride is metabolised in the liver to an epoxide that causes the cancer. As well as a Maximum Exposure Limit, a yearly exposure limit of 3 ppm applies to exposed workers. High levels of chronic liver disease have been reported in refrigeration engineers, chemists, dry cleaners, rubber manufacturers, and workers exposed to carbon tetrachloride and plutonium.

The Kidney and Urinary Tract

Proven or suspected kidney toxins (nephrotoxins) include arsenic, beryllium, lead, cadmium, mercury and uranium plus their compounds, solvents, and pesticides. Kidney failure from exposure to lead was common earlier this century, and kidney disease is the best known effect of chronic exposure to cadmium.

The kidneys, via urine, are the major route by which toxic chemicals are excreted from the body. Because of this, and the way the kidneys do their job, they are vulnerable to the toxic effects of chemicals. The damage caused is complicated, and in many cases still not well understood, but can result from acute and chronic exposure.

According to the US National Bladder Cancer Study, up to a quarter of bladder cancers are caused by work. Aromatic amines are one group of chemicals known to cause bladder cancer. Occupationally-induced bladder cancer was first reported in 1895 in German dye-manufacturing workers.

Because they use dyes, jobs where workers are at risk of developing bladder cancer include textile, fur and leather dyeing as well as aromatic amine manufacturing. Aromatic amines are also used in rubber and plastics manufacture. Even though bladder cancer usually takes around 20 years to develop after exposure to aromatic amines, this can range from 4-40 years, and can result from exposures as short as 19 weeks (Levy and Wegman).

MbOCA is one aromatic amine used as a stabiliser in plastics manufacturing, often by small manufacturing plants. After an HSE national enforcement project in 1996-97, the HSE described conditions in many of these workplaces as "rough and ready." MbOCA is easily absorbed through the skin, and in many of these small factories even the eating areas and maintenance workers' tools were contaminated.

Blood

Chemicals which affect the blood can do so either by being directly toxic to blood cells, or by preventing it from delivering oxygen to the rest of the body. Haemoglobin is the molecule which carries oxygen to the body's tissues, and by combining with it to form carboxyhaemoglobin, carbon monoxide prevents oxygen reaching the tissues. Carbon monoxide is produced when organic substances such as gas and petrol are incompletely burnt, and as it is present in car exhausts, fire fighters, garage workers and traffic police can be exposed to high concentrations. Carbon monoxide is also produced by poorly maintained gas fires, and can also be produced in the body by metabolism of some chemicals like methylene chloride. Blood carboxyhaemoglobin levels are around 0.5 per cent in non-smokers, and 5 per cent in smokers. Headaches occur at levels of 10-20 per cent, and a level of 60-70 per cent is fatal within hours.

Other substances combine with haemoglobin to form methaemoglobin, which is also unable to deliver oxygen to the tissues. Workplace chemicals which have this effect include aniline dyes, nitrous gases, potassium chlorate, nitrobenzenes, phenylenediamine, and toluenediamine. Methaemoglobinaemia also results from environmental exposures to foods high in nitrates or nitrites, or well water contaminated with nitrates (Levy and Wegman). As well as binding with red blood cells to interfere with the carriage of oxygen, other chemicals can break up red blood cells. This is known as haemolysis, and is caused by chemicals like naphthalene, copper, organic compounds of metals such as tributyltin, and arsine. Arsine gas is an inorganic arsenic compound, and although it is used in the electronics industry, it is most dangerous because it can be formed accidentally when arsenic-contaminated metals or coal come into contact with acids. There have been cases of sewer workers being gassed by arsine after using acids to clear drains previously contaminated with substances containing arsenic. As well as its effects on red blood cells, arsine also causes kidney failure.

Lead also affects the blood by interfering with red blood cell manufacture. These cells are produced in lower quantities and do not live as long as normal, causing anaemia. Workers exposed to lead, such as welders, painters, jewellers, and in smelters, foundries and potteries, are covered by special legislation in the UK and should have biological monitoring to assess their exposure. Benzene is another well known example of a chemical which causes blood toxicity. It is metabolised in the body to a chemical which damages bone marrow, the site where blood cells are produced. Leukaemia, a form of cancer, was first linked with exposure to benzene in the 1920s. It also causes aplastic anaemia which, like leukaemia, is often fatal.

The Heart and Blood Vessels

Much less work has been done on occupational, as opposed to hereditary or life-style, factors associated with heart disease. However, because heart disease is the largest cause of death among both men and women in the UK, even a small reduction in risk due to occupational exposure could involve large numbers of people, and be an important public health measure. There is good evidence that occupational exposure to certain materials, such as the solvent carbon disulphide, is linked with heart disease.

Explosives manufacturing workers exposed to nitroglycerin and ethylene glycol dinitrate can suffer from angina when away from work, because these chemicals (like those used to treat angina) cause the heart's blood vessels to expand. Acute exposure to some solvents has also been associated with sudden death, probably due to changes in the heart's rhythm.

The Reproductive System

Reproductive hazards can be divided into two groups, those which cause reproductive effects, such as chemicals which impair sperm production and fertility, and those which damage the developing foetus. Reproductive hazards affect both men and women at work. Several chemicals are known to make men less fertile by affecting the production of sperm. Other substances can reduce sex drive or cause impotence.

In women, some occupational exposures are associated with irregular periods, increased rates of miscarriage, or premature or low birth weight babies (which are less healthy and more likely to die before their first birthday than other babies), or deformed babies.

Chemicals that cause birth defects are known as teratogens, and the effects they have as 'teratogenic'. These can include visible (structural) deformities or functional abnormalities, like learning difficulties.

As well as the dose a woman is exposed to, the stage of her pregnancy at which the exposure takes place is important. The developing foetus is very vulnerable to teratogens during the first three months of pregnancy, precisely the time when a woman may not realise she is pregnant. Because of this, reproductive hazards at work should be identified before women become pregnant. Only about 4 per cent of the chemicals in commercial use in the USA have been tested for teratogenicity. But of those tested on animals 37 per cent are clearly, probably or possibly teratogenic. Reproductive effects have been reported in many occupational groups, including health care workers, laboratory and dental technicians, factory workers, pulp and paper industry workers, construction workers, transport and communication workers, printers, plastics industry workers, and lead production workers.

Table. Male Reproductive Hazards Reported in Human and Animal Studies

effect	in men exposed to	in animals exposed to
impotence or low sex drive	metals (lead, manganese, mercury), toluene di-isocyanate, vinyl chloride, chloroprene	
testicle damage or infertility	pesticides [Kepone, dibromochloropropane (DBCP)], chloroprene, lead	benzene, benzopyrene, boron, cadmium, epichlorohydrin, ethylene dibromide, polybrominated biphenyls
toxicity to sperm	pesticides (carbaryl, DBCP), carbon disulphide, lead, radiation, heat stress, toluenediamine + dinitrotoluene, cytotoxic drugs	arsenic, chloroprene, ethylene glycol ethers, ethylene oxide, halothane, Kepone, mercury, nitrous oxide, trichloroethylene, triethyleneamine

In the UK and the rest of Europe, safety data sheets and labels should contain information on reproductive hazards, including the risk phrases "R46 (may cause heritable genetic damage)", "R61 (may cause harm to the unborn child)", "R63 (possible risk of harm to the unborn child)" and "R64 (may cause harm to breast-fed babies)" required by the CHIP Regulations.

Table. Reproductive Hazards in Women Reported in Human and Animal Studies

effect	in women exposed to	in animals exposed to
irregular periods and other gynaecological disorders	aniline, benzene, chloroprene, formaldehyde, inorganic mercury, polychlorinated biphenyls (PCBs), styrene, toluene	
abortion or infertility	anaesthetic gases, aniline, arsenic, benzene, ethylene oxide, cytotoxic drugs, ethylene oxide, formaldehyde, lead, 2,4,5-trichlorophenol	
foetal toxicity or death		chloroform, dichloromethane, ethylene dichloride, inorganic mercury, nitrogen dioxide, polybrominated biphenyls, selenium, tetrachloroethylene, thallium, trichloroethylene, vinylidene chloride
low birth weight	carbon monoxide, formaldehyde, PCBs, toluene, vinyl chloride	
premature birth	lead, heat stress	
teratogenicity	hexachloroprene, radiation, organic mercury, vinyl chloride	arsenic, benzopyrene, chlorodifluoromethane, chloroprene, monomethyl formamide, acrylonitrile, methyl ethyl ketone, tellurium
cancer	diethylstilboestrol (DES), hepatitis B	arsenic, benzopyrene, vinyl chloride

The results of two large studies on rates of miscarriage (spontaneous abortion) were published in 1997. One, funded by the HSE, looked at dry cleaning workers in the UK. The other examined all previously published research on women exposed to anaesthetic gases at work. Although perchloroethylene has been used as an industrial solvent for over 50 years, and in dry cleaning for more than 30 years, concern over its effects on pregnant women arose after a Finnish study in 1980 which found dry cleaning workers twice as likely to miscarry as other women. The UK study looked at 7305 women who worked in dry cleaners, and found they were half as likely again as other women to report that they had a miscarriage. A similarly increased risk of miscarriage was found for women exposed to anaesthetic gases.

CANCER

Several carcinogens (cancer-causing agents) have already been mentioned. However, it is more usual to discuss cancer-causing chemicals together as a group, irrespective of which organs or systems they affect. Cancer is not one, but dozens of diseases affecting different organs and tissues. Cancer kills over one in five of the population of industrialised countries, due to both inherited and environmental (including diet and life-style) factors as well as workplace exposures. How much cancer results from workplace exposures has been argued about for years. In 1978, scientists from the US National Cancer Institute, National Institute of Occupational Safety and Health, and National Institute of Environmental Health Sciences estimated 20-40 per cent of cancers were work-related. Two UK scientists, Richard Doll and Julian Peto, said in 1981 that only 2-8 per cent of all cancers were due to work. These figures have been debated ever since, with arguments from a variety of standpoints.

A 1996 review of occupational lung cancers said that 9,000-10,000 men and 900-1,900 women develop lung cancer each year in the USA due to past occupational exposure to carcinogens. This represents 9 per cent of male and 2 per cent of female US lung cancer cases. Others say that 17 per cent of US male lung cancers stem from exposure to carcinogens at work.

Cancers are complex diseases, because they usually result from several factors, and often take many years to develop. There are also several different steps involved in the process which turns normal cells into cancerous tumours. As a result, carcinogenesis (the development of cancer) is described as a multi-step process. The two most important stages are initiation and promotion. Some chemicals or their metabolites act as initiators, causing permanent changes in a cell's genetic makeup (mutation). Others will then promote the development of these abnormal cells into a tumour. Some chemicals act as both initiator and promoter. However, not all chemicals that cause mutations (mutagens) are carcinogens, and some chemicals cause cancer without damaging DNA.

Synergism

The risk of developing lung cancer due to smoking and/or exposure to asbestos is a useful example of synergy, which occurs when the combined effect of two chemicals is much greater than their additive effects. Smokers have a ten-times greater risk of dying from lung cancer than non-smokers, and if people are exposed to asbestos, they are five times more likely to die of lung cancer compared with those not exposed to asbestos. However, the combined risk of dying of lung cancer in a smoker also exposed to asbestos is not 15 times greater than a non-smoker not exposed to asbestos, but 80 times greater.

Testing chemicals for their carcinogenicity is expensive, but two organisations which conduct a lot of carcinogenicity testing are the National Toxicology Programme (NTP), part of the US government National Institutes of Health, and the International Agency for Research on Cancer (IARC), part of the World Health Organization.

Since 1972, IARC has published over 70 monographs reviewing the scientific evidence for carcinogenicity of individual chemicals, as well as mixtures and specific occupations. By 1998, about 834 chemicals, groups of chemicals, complex mixtures, and occupational exposures had been evaluated.

Of these, IARC says 75 are carcinogenic to humans (group 1), 59 are probably carcinogenic to humans (group 2A), 225 are possibly carcinogenic to humans (group 2B), and the rest are unclassifiable (group 3). IARC classifications are based on the strength of the scientific evidence, not the potency of the carcinogen.

"Keep in mind that a chemical can be demonstrated to be a human carcinogen only if an opportunity exists to study it in exposed humans in a systematic way, and such opportunities are not frequently found".

"Occupational cancer differs from other occupational diseases in several ways: no safe level of exposure to carcinogens is recognised; many different forms of cancer exist; cancer develops many years after exposure; occupational cancer generally resembles cancer of non-occupational origin; and competing carcinogenic exposures are present in many cases.

On the other hand, occupational cancer shares at least important features with other occupational diseases: there are large data gaps in relating exposure to disease; and most cases are preventable".

The Immune System, Allergies and Sensitization

The body's defence against foreign substances and organisms is known as the immune system, and immune reactions play an important part in many work-related diseases. The immune system is complex and relies on the work of many different types of cells which are found in bone marrow, thymus, spleen, lymph nodes, in the lining of the gut and respiratory tract, and in the skin.

The action of some cells is very general, but others respond to specific substances. One of the hallmarks of this specific immunity is that "memory" develops, so that when the body comes into contact with the same substance a second time the response is faster and stronger. Specific immunity is divided into antibody-mediated and cell-mediated systems, and substances that can cause immune responses are called antigens; many of them are proteins. Hypersensitivity or allergic reactions can be classified into four distinct types: Type I, anaphylactic or immediate hypersensitivity reactions in which symptoms occur within minutes of exposure, Type II, cytotoxic reactions in which symptoms appear within hours, Type III, immune complex reaction, and Type IV, cellular immunity or delayed-type hypersensitivity in which symptoms develop 24-48 hours after exposure.

The most common allergies at work affect the skin and respiratory tract. Common causes of allergic contact dermatitis include latex (in surgical gloves), formaldehyde, rubber additives, nickel or chromium compounds, bactericides and fungicides, adhesives and sealants, plants and wood, and hairdressing chemicals. Common causes of allergic asthma and rhinitis include latex, flour, mites, enzymes, di-isocyanates, metals and metal compounds, and certain drugs.

Multiple Chemical Sensitivity

Multiple chemical sensitivity (MCS) is the name given to a very wide range of symptoms, related to a large number of chemicals. MCS has been called by various names, including "environmental illness" and, most recently, "toxicant-induced loss of tolerance" (TILT). MCS has been much talked about since the 1980s but research so far has not yet uncovered the underlying mechanisms.

Some doctors believe that the symptoms of MCS sufferers are "all in the head." This is a view which is far from helpful for patients. Dr Joseph LaDou, Professor of Occupational Medicine at the University of California says, "Although the aetiology of MCS is controversial, the patient may be suffering from disabling symptoms, frustrated by the lack of definitive answers from clinicians, and is sometimes desperately seeking advice and counsel regarding treatment. Approaching the history with the suspicion that the patient with MCS is suffering from a psychiatric disorder, is malingering, or seeking monetary benefits is not helpful."

Although a psychological cause for MCS has not been ruled out, evidence from most research on MCS points to a physical cause or causes. What puzzles doctors, and frustrates sufferers, is that the symptoms vary widely between individuals, and the substances associated with it are also numerous. Although there are several definitions, most agree that MCS affects one or many organ systems, symptoms come and go, and symptoms are brought on by very low exposures to many different chemicals.

Although they vary a great deal, the symptoms of MCS often involve the upper respiratory tract (blocked nose, dryness or burning), central nervous system (problems with memory or concentration, insomnia, drowsiness, irritability or depression), and gut, as well as muscle or joint pain, headaches and tiredness. These symptoms are triggered by low-level exposure to a large range of substances, but the initial cause will usually have been exposure to one of a smaller number of initiators. In other words, MCS is a two-stage process, involving initiation and triggering.

Table. Initiators and Triggers of MCS

Common initiators	Pesticides (particularly organophosphates and carbamates), solvents, carpets and glue, mercury amalgam, formaldehyde
Common triggers	Air fresheners, alcohol, car exhaust, cleaners/detergents, cosmetics, foods, nail varnish, newly painted rooms, newspapers/printed material, perfumes, solvents, tobacco smoke.

Researching MCS

In their book on multiple chemical sensitivity (MCS), Drs Nicholas Ashford and Claudia Miller say, "In recent years, we have observed a tendency to name MCS-like conditions after the suspected initiating event, for example, darkroom disease and Gulf War syndrome.

This may cloak a larger view that there is an underlying, unifying mechanism, for example, that some people lose tolerance following certain chemical exposures, and that thereafter their symptoms are triggered (and their illness is perpetuated) by common, low-level exposures."

Exposures that appear to initiate MCS are most often pesticides (especially organophosphates) and solvents. However, once the illness has a foothold, triggers are often substances like perfumes, tobacco smoke or cleaning agents. Because of this, patients often mistake triggers for the initiator.

They cite recent US research, involving random telephone surveys, which found that around 5 per cent of people reported symptoms of MCS. In other surveys, 15-34 per cent said they were unusually sensitive to certain chemicals. Although there are now many theories about the mechanisms involved in MCS, much more research is needed. According to Ashford and Miller: "Funding agencies will need to make a much greater financial commitment if progress is to be made. By not understanding the causes of chemical sensitivity, we take an immense gamble - but knowledge will not come cheaply. Understanding chemical sensitivity is pivotal to establishing sound environmental policy. If there is a subset of the population that is (or can become) especially sensitive to low-level chemical exposures, a strategy for protecting them must be found".

Fire, Explosion and Radiation

Many chemicals used at work are flammable or explosive, either on their own or when mixed together. Dust explosions can also occur if enough dust of any solid material in the air is ignited by a flame or spark. Risks of fire and explosion need to be controlled by storing and handling chemicals correctly. Certain chemicals emit ionising radiation, a specific type of radiation in the electromagnetic spectrum (other types include radiowaves and visible light). Some kinds of ionising radiation penetrate body tissues. High doses of radiation, as occurred after the dropping of atomic bombs on Hiroshima and Nagasaki, and in Chernobyl after the nuclear reactor accident, are lethal. At low doses, ionising radiation causes DNA damage and so can lead to cancer. Workers at risk of exposure to ionising radiation include those in hospitals and laboratories.

The Ionising Radiation Regulations 1985 require that exposures are kept "as low as is reasonably practicable". Exposure above specified single and annual doses must be reported to the HSE. The regulations also specify that certain areas be designated as "controlled areas" or "supervised areas" according to likely radiation exposure.

Employers must also appoint qualified radiation protection advisers in workplaces with "controlled areas" or where doses above a certain level occur. New regulations to implement two Euratom directives will come into force during 2000. A European directive on physical agents, which would have covered non-ionising radiation, appears unlikely to see the light of day.

TOXICITY TESTING

Understanding how chemicals are tested for toxicity is very important for trade union safety reps and members as users. The results of these tests often appear in safety data sheets, they are used to decide how chemicals are used, controlled, labelled, and, most importantly, in setting occupational exposure limits.

Newly introduced chemicals are now required to have a certain set of toxicity data, but these do not exist for many chemicals already in use before legislation was introduced. It is important to remember that absence of evidence of risk is not the same thing as evidence of absence of risk.

In 1992, the United Nations Conference on Environment and Development called on member countries and the Organisation for Economic Co-operation and Development to complete toxicity testing for 2,550 industrial chemicals produced in high volumes (at least 1,000 tonnes per year in any member country). In 1998 the OECD announced it had completed 109 tests.

How much (or how little) is known about the toxicity of even the most widely used chemicals sparked a major debate in the US press during 1997, between the Environmental Defence Fund (EDF) and the Chemical Manufacturers' Association (CMA). The EDF said that toxicity data was inadequate for 71 per cent of the 3,000 chemicals made and used in the USA in the highest amounts.

The CMA disagreed, saying the figure was only 53 per cent, but only because they included industrial data that had not been made publicly available. The US Environmental Protection Agency said, "There is a problem with public availability of basic screening information on chemicals." To try and speed up this voluntary approach to testing, the EDF mounted a 'naming and shaming' campaign. They asked the top 100 chemical companies if they would find and disclose basic toxicity data on the high-volume chemicals they produced by January 2000. The names of those which would not commit themselves were then listed in US newspaper advertisements.

According to UK academic Professor Andrew Watterson of the Centre for Occupational and Environmental Health, "The demands of carrying out complete health assessments on the tens of thousands of chemicals, metals and other substances used in the world today are beyond the resources and abilities of the global scientific community." Instead, he argues for a precautionary approach such as toxics use reduction.

Testing for the Effects of Acute Exposure

Toxicity studies can be divided into laboratory studies using live animals (*in vivo* studies) or groups of cells (*in vitro* studies), and studies of human populations (epidemiological studies). Animal studies are used to test chemicals for their acute and chronic toxicity by various routes of exposure.

The standard way of measuring a chemical's acute toxicity is to feed it at a range of single doses to groups of laboratory animals, such as rats. The dose that kills 50 per cent of the group, the LD50 (lethal dose-50), is then recorded as well as the effects noticed in the animals. One use of LD50 values is in deciding how to label chemicals under the Chemicals (Hazard Information and Packaging for Supply) (CHIP) Regulations.

Table. Examples of Toxicity Measures

Category	**LD50 oral rat mg/kg**
very toxic	Less than 25
Toxic	From 25 to 200
Harmful	From 200 to 2000

Acute toxicity of methyl ethyl ketone (MEK)

LD50 oral rat	**2,740 mg/kg**
LD50 dermal (skin) rabbit	13 g/kg
LC50 (2 hour) inhalation mouse	40 g/m3
LDLO intraperitoneal (abdominal) guinea pig	2000 mg/kg

Table lists various measures of the acute toxicity of MEK, and illustrates some of the differences that are used in these tests. Table lists four routes of exposure, oral, skin, inhalation and intraperitoneal injection, and four of the most commonly used laboratory species.

Inhalation toxicity is described as a lethal concentration-50 (LC50) rather than LD50, and concentrations are expressed as milligrams or grams of substance per m^3 of air, or in parts per million (ppm). The LC50 is only meaningful if the duration of exposure is known. The other doses are expressed as milligrams (mg) or grams (g) of substance per kilogram (kg) of body weight.

A chemical's irritancy is tested by applying it to the eyes and skin of animals. One such test is known as the Draize test, which involves applying chemicals to rabbits' skin and eyes, rabbits being chosen because of their large and exposed eyeballs. The Draize test has frequently been criticised by animal rights campaigners, particularly when used to assess the irritancy of cosmetic products.

Testing for the Effects of Chronic Exposure

The effects of chronic exposure are also tested in animals. In these tests, the animals are fed, inhale, or have the chemical painted on their skins throughout their lives. The type and amount of disease they develop are then compared with the effects in a control group. The controls should be the same as the exposed group in every respect except the chemical exposure.

They should, for example, be the same strain of the same species, and be fed the same diet. The differences in rates of various diseases are then tested by various statistical means to see if they are significant.

Having an adequate control group is just one element of good experimental design, as is the number of animals used. If a study is badly designed, its results will not be reliable. For example, when testing carcinogens on animals the US National Cancer Institute says that the chemical should be tested at two doses in both sexes of two species of rodent, and each group should contain at least 50 animals. One dose is often the highest dose that will not kill, or acutely poison, the animal.

Although this means that positive results will not be missed just because a high enough dose was not used, it means that predicting the effects of very low levels of the chemical is difficult. Trying to extrapolate (predict) what will happen at low doses, or what will happen in humans, are two of the major problems of animal testing.

Dose-response

When the results of these tests are plotted on a graph, with dose along the bottom (horizontal axis) and response up the side (vertical axis), a dose-response curve is obtained. (It is still called a curve even if it is a straight line!)

For example, the dose could be milligrams of chemical per kilogram of body weight per day (mg/kg/day) and the response could be the percentage of animals that got a certain type of cancer, or suffered a certain degree of liver damage. The question of whether there is a threshold dose, below which there is no toxic effect (no observed effect level, or NOEL) is controversial, especially for carcinogens. Because many carcinogens act by damaging DNA, and once this damage has been done it is permanent and by definition results in an increased risk of cancer, many argue that there is no such thing as a safe dose of these carcinogens. Also, because cancer is such a serious disease, and so many gaps exist in our knowledge of how carcinogens act, it makes sense to take a precautionary approach. In the Carcinogens Approved Code of Practice the HSE says, "the risk of cancer from exposure to a substance cannot in most cases be presumed to be zero except by eliminating exposure." NOELs are also used in setting occupational exposure limits.

"We assume that every molecule of benzene to which an individual is exposed has some finite risk, albeit small, of producing a mutation that may result in acute myelogenous leukaemia".

Usefullness of Animal Studies

Some researchers in Canada and the USA have done interesting work comparing the attitudes of toxicologists and the public towards chemical risks. The public is more likely than toxicologists to think chemicals pose greater risks, and also finds it difficult to understand the concept of dose-response relationships. The public is much more likely than toxicologists to think the results of animal carcinogenicity studies can be applied to humans.

The study also found much disagreement between toxicologists about how to interpret various results. No wonder the public is confused, when the US study says, "Among the most important findings in this study was the high percentage of toxicologists who doubted the validity of the animal and bacterial studies that form the backbone of their science." Fewer toxicologists in industry than in university or government jobs agreed that animal carcinogens could reasonably be expected to cause cancer in humans.

"Good toxicology recognises its limits, its gaps in knowledge and the debates about the validity of its understanding of the mechanisms of toxicity in the context of moving towards practical public health precautionary policies on toxics use in our workplaces,".

Testing for Carcinogens

As well as long-term animal tests, a range of other tests are used to predict whether a chemical is likely to cause cancer. These are done with cell cultures, for example using mouse lymphoma cells, and micro-organisms such as the bacteria *Salmonella* (the Ames test) and *Escherichia coli*, and are known as *in vitro* (literally, in glass) tests. They are quicker, cheaper, and more ethically acceptable than animal tests.

They test a chemical's ability to damage genetic material (genotoxicity), although as mentioned before, most but not all genotoxins cause cancer, and (more importantly) not all carcinogens act by damaging genes.

These would not, therefore, be detected if only *in vitro* tests were used. While animal and *in vitro* testing might provide regulators with a lot of data, because the mechanisms leading from exposure to effect are often unknown, there is still much educated guesswork involved in risk assessment.

Genetic Toxicology Tests

Tests for gene mutations in bacteria	*Salmonella*
	Escherichia coli
Tests for gene and chromosome mutations in mammal cells	Chinese hamster ovary (CHO) cell assay
	V79 Chinese hamster lung fibroblast assay
	L5178Y/$tk^{+/-}$ mouse lymphoma cell assay
Tests for chromosome mutations	Chromosome aberration in CHO cells *in vitro*
	Chromosome aberration in human lymphocytes *in vitro*
	Micronucleus test in erythrocytes from mouse bone marrow *in vivo*
	dominant lethal test in rodents
Tests to measure repair of DNA damage	sister chromatid exchange *in vitro* or *in vivo*
	Unscheduled DNA synthesis *in vitro*

EPIDEMIOLOGY

The other way of studying the toxic effects of chemicals is to look at patterns of disease in human populations, such as groups of exposed workers. These are obviously more relevant, because they involve humans rather than animals, but can be difficult to interpret.

They can also only be done after human exposure has occurred. But they do allow populations to be followed who have been exposed to chemicals cleared by toxicologists. The main types of epidemiological study are case-control studies and cohort studies.

Case-control studies compare a group of people with a specific disease with another (control) group who do not have the disease but are the same in other respects, like sex, age and social class. Information is then collected on both groups' occupational exposures.

Because neither their doctors nor employers are likely to have adequate records of which chemicals workers have been exposed to, and in what quantities, epidemiologists often rely on people's memories. The results are expressed as an odds-ratio (OR). If the OR is significantly greater than 1.0, there could be a link between the exposure and the disease.

Cohort studies look at a group known to be exposed to a particular chemical, and compare their death rate or diseases with a similar, unexposed, group. The results are expressed as Relative Risk (RR). An RR greater than 1.0 suggests a link between the exposure and the disease(s). If a control group cannot be found, comparisons are made between the exposed group and the general population, and expressed as a standardised mortality ratio (SMR).

Epidemiologists weigh the evidence by looking at results from as many studies as possible, as well as the size of the OR RR, the existence of a dose-response relationship, and biological plausibility (is there a believable basis for the exposure leading to the disease?). A scientific consensus may then be reached, as happens with the IARC cancer evaluations.

Healthy Worker Effect

The results of epidemiological studies obviously depend on the comparisons made between the cases and control groups used. If the groups are not well matched, the results will not be meaningful. For this reason cases and control groups may be matched for age, sex and race, as well as lifestyle factors like smoking and alcohol consumption. Occupational groups very often have lower total mortality than the general population as the latter includes people unable to work due to illness or disability. In other words, any group of workers is likely to be more healthy than the population as a whole, a phenomenon known as the "healthy worker effect".

Other Factors

Different people can respond very differently to the same dose of a chemical, and this variation is due to many factors. Age, sex, and genetics all influence how susceptible people are to chemicals. People can also be more at risk from toxic chemicals if taking certain drugs, if they have certain diseases, and if they are exposed to mixtures of chemicals. The very young and the old are usually more susceptible to toxic chemicals than the rest of the population. In the young, this is because they can absorb relatively large doses of chemicals, the systems that detoxify chemicals have not yet matured, and parts of their body, such as their brains, are more easily damaged.

Older people are more susceptible because they may have larger body stores of certain chemicals, and more of them have diseases which could put them at greater risk from toxic chemicals. In addition to reproductive and developmental factors, men and women may differ in their response to chemicals in other ways. As women in general have lower body weight than men, they will be more strongly affected by the same dose of chemical. Different genetic makeup and different proportions of body fat may also have an influence.

For example, as women have a higher proportion of body fat than men, they are more prone than men to retain organochlorine pesticides such as DDT once ingested. There are reports that women are more susceptible to skin diseases induced by chemicals and to the effects of indoor air pollution including multiple chemical sensitivity. The possibility of differential responses should be taken into account when performing risk assessments.

Workers' Epidemiology

Just because a scientific consensus does not exist, or research has not been done, does not mean that workers and safety reps should not act. Increasing numbers of workers are doing their own research, usually helped by their trade unions. This is sometimes called workers' epidemiology, lay epidemiology or participatory research. It might involve gathering and analysing information by questionnaire, or pinpointing health problems by body mapping.

Getting useful information from a questionnaire depends on asking the right questions, and the best way of doing this is by developing the questionnaire within the group of workers concerned. There are, however, several standard questionnaires available to survey particular workplace health problems, which could be used as the basis of such a questionnaire.

The GPMU, for example, is currently involved in piloting a questionnaire to measure the rate of occupational skin disease among printers in the Nottingham area. But gathering data is only the first step, the information will need to be analysed (cheap computer packages are available to do this) and acted upon! "One of best sources of information on occupational hazards is one of the most frequently overlooked - exposed workers".

Body Mapping

Another method of spotting new or underestimated health problems at work is body mapping, which can also be organised by safety reps. The process involves getting a group of workers together who do the same job. Then, using two large maps or outlines of the body (one for the front and one for the back), workers mark (with coloured stickers or pens) areas of the body affected by their work. As they do so, they should give the group a brief explanation of what they have marked, and why, and these comments can be written around the edge of the map. Hazards magazine produces packs of body maps and has published information on body mapping and workers' epidemiology.

HARMFUL CHEMICALS IN OUR ENVIRONMENT

Over the past century humans have introduced a large number of chemical substances into the environment. Some are the waste from industrial and agricultural processes. Some have been designed as structural materials and others have been designed to perform various functions such as healing the sick or killing pests and weeds.

Obviously some chemicals are useful but many are toxic and their harm to the environment and our health far outweighs their benefit to society. We need to manage the risks better by only using chemicals, which are safe.

Chemicals enter air as emissions and water as effluent. Industrial and motor vehicle emissions of nitrogen and sulphur oxides cause acid rain, which poisons fish and other aquatic organisms in rivers and lakes and affects the ability of soil to support plants. Carbon dioxide causes the greenhouse effect and climate change. C hlorofluorocarbons (CFCs) cause the destruction of ozone in the stratosphere and create the possibility of serious environmental damage from ultraviolet radiation. Chemical fertilizers and nutrients run-off from farms and gardens cause the build up of toxic algae in rivers, making them uninhabitable to aquatic organisms and unpleasant for humans. Some toxic chemicals find their way from landfill waste sites into our groundwater, rivers and oceans and induce genetic changes that compromise the ability of life to reproduce and survive.

The impact of human activities on the environment is complex and affects a chain of interconnecting ecosystems. The extinction of species all along the chain may mean the loss of useful genetic material or life saving cancer drugs or safer alternatives to the dangerous chemicals in use at the moment.

ORGANOCHLORINES

Organochlorine compounds such as polychlorinated biphenyls or PCBs were developed originally for use in electric equipment as cooling agents and are very dangerous chemicals. During the manufacture and disposal of products containing PCBs, and as a result of accidents, millions of gallons of PCB oil have leaked out. Although their manufacture in the United States was halted in the 1970s and they are being phased out, they are difficult to detect, are nearly indestructible and large quantities remain in existence and they will remain in the environment for a long time.

They accumulate in the food chain and significant levels of them have been found in marine species, particularly mammals and sea birds, decades after their production was discontinued. They are carcinogenic and capable of damaging the liver, nervous system and the reproductive system in adults. When PCBs are burned, even more toxic dioxins are formed.

DIOXINS

Dioxins, are a class of super-toxic chemicals formed as a by-product of the manufacture, moulding, or burning of organic chemicals and plastics that contain chlorine. They are the most toxic man-made organic chemicals known.

They cause serious health effects even at levels as low as a few parts per trillion. Only radioactive waste is more toxic. They are virtually indestructible and are excreted by the body extremely slowly. Dioxins became known when Vietnam War veterans and Vietnamese civilians, exposed to dioxin-contaminated Agent Orange, became ill.

Dioxins enter the body in food and accumulate in body fat. They bind to cell receptors and disrupt hormone functions in the body and they also affect gene functions. Our bodies have no defence against dioxins which may cause a wide range of problems, from cancer to reduced immunity to nervous system disorders to miscarriages and birth deformity. The effects can be very obvious or subtle. Because they change gene functions, they can cause genetic diseases to appear and they can interfere with child development. Attention Deficit Disorder, diabetes, endometriosis, chronic fatigue syndrome, rare nervous and blood disorders have been linked to exposure to dioxins and PCBs.

Over the past 40 years there has been a dramatic increase in the manufacture and use of chlorinated organic chemicals in plastics, insecticides and herbicides. Dioxins have been found in high concentrations near to the sites where these chemicals have been produced and where insecticides and herbicides have been heavily used, such as on farms, orchards, or along electric and railway lines. They have also been a found downstream from paper mills where chlorine chemicals have been used to bleach wood pulp. In the last few years we have begun to discard our unfashionable household plastic products, together with industrial and medical waste by burning them in incinerators. Dioxins formed during the combustion process have been carried for hundreds of miles on tiny specks of ash and contaminated the countryside.

They settle on pastures and crops and get eaten by cows, pigs and chickens. They get into lakes, streams, and ocean and are taken up by fish. They go through the food chain and appear in meat and milk and accumulate in the fat cells of our bodies.

CADMIUM

Cadmium occurs naturally in the earth's crust combined with other elements. It is usually formed as a mineral such as cadmium oxide, cadmium chloride, or cadmium sulphate and although these compounds are highly toxic they are less harmful when bound to rocks. They are present in coal and in the soil. Cadmium is useful because it doesn't corrode easily. It is used in batteries, plastics, pigments and metal coatings. Cadmium gets into the environment through landfills, poor waste disposal methods and leaks at hazardous waste sites. It is produced by mining and other industrial activities.

Cadmium particles enter our air when we burn coal for energy and incinerate household waste. The particles can travel far before falling to the ground or water. Each year many tonnes of cadmium are discharged into our seas and oceans. Animals and plants take up cadmium when it is in the environment. If we consume food contaminated with cadmium it can irritate our digestive system and cause vomiting and diarrhoea. If inhaled it can damage our lungs. Even when levels of exposure are low, over time, cadmium accumulates in the body and it can be difficult to get rid of. Accumulated cadmium can cause kidneys and bone disease.

We take cadmium into our body by:

- Inhaling it when working in factories that make batteries or do welding, brazing or soldering.

- Inhaling it when near power stations or factories burning of fossil fuels.
- Eating foods in which it accumulates such as shellfish, liver and kidney.
- Drinking water that is contaminated.
- Smoking cigarettes.

Potentially dangerous chemicals such as these are being introduced into the environment all the time. As in the case of PCBs their effect on living things may not be known until many years after their release. Hundreds of thousands of different chemicals are marketed worldwide. Of these 5000 are produced in quantities over 10 tonnes a year and 1500 are produced in quantities over 1000 tonnes year. We do not have enough information about the environmental effects of these industrial chemicals and their effects on humans. The balance between human activity and ecological sustainability is wrong.

Ways to Protect Environment

Use biodegradable products. Make your own cleaning agent using safe materials. Dispose of chemical waste carefully. Do not put them down the sink. Be wise with home maintenance and in the garden. Do not burn plastics.

Avoid all organic chemicals that have "chloro" as part of their names including wood preservatives, herbicides and insecticides. Avoid chlorine bleach (sodium hypochlorite) and products containing it. Use oxygen bleach instead. Use unbleached paper products.

Avoid "Permethrin" flea sprays for pets. Avoid products made of or packaged in polyvinyl chloride (PVC). Avoid cling flim plastic wraps unless they are clearly identified as non-chlorinated plastic.

To minimise your risk of dioxins accumulating in your body avoid all full-fat dairy products and fatty meats such as beef or pork. Wash all fruits and vegetables to remove chlorophenol pesticide residue.

Avoid grapes and raisins unless they are clearly labelled as organically grown. Avoid soaps, toothpaste and deodourants containing "triclosan," a chlorophenol.

We can reduce the dioxins if we stop producing PVCs and other chlorinated chemicals. If your local government sends its waste to an incinerator, request that they stop burning plastics and introduce a comprehensive recycling service. Write to companies and ask them to use safe substitutes to chlorinated plastics. Ask your supermarket to sell Totally Chlorine Free (TCF) products. Join or form a local environmental group campaigning against hazardous chemicals. People who work with cadmium should take care not to inhale cadmium-containing dust and should avoid carrying it home from work on their clothes, skin or hair. Eat from a wide range of foods to prevent the risk of ingesting toxic levels of cadmium.

HAZARDOUS CHEMICALS

Chemicals, for which there is statistically evidence based on at least one study conducted in accordance with established scientific principles that acute or chronic health effects may occur in exposed employees, is classifies as a health hazard. Health hazards include chemicals that are carcinogens, toxin or highly toxic agents, reproductive toxins, irritants, corrosives, sensitizers, hepatotoxins, nephrotoxins, neurotoxins, agents which act on the hematopietic systems; and agents that damage the lungs, skin, eye or mucous membranes are hazardous chemicals. Many chemicals can cause harm by virtue of their toxicity.

In this case, the toxicity of a chemical refers to its ability to damage an organ system, such as the liver and the kidney or to disrupt a biochemical process such as the blood clotting mechanism or to disturb an enzyme system at some sites in the body removed from the site of contact. In actual practice, hazards do not group themselves in neat categories, but usually occur in combination and/or sequence. The following are some characteristics of flammability, a common chemical hazard.

MERCURY

Exposure to mercury occurs from breathing contaminated air, ingesting contaminated water and food, and having dental and medical treatments. Mercury, at high levels, may damage the brain, kidneys, and developing fetus.

when Mercury Enters the Environment

- Inorganic mercury (metallic mercury and inorganic mercury compounds) enters the air from mining ore deposits, burning coal and waste, and from manufacturing plants.
- It enters the water or soil from natural deposits, disposal of wastes, and volcanic activity.
- Methylmercury may be formed in water and soil by small organisms called bacteria.
- Methylmercury builds up in the tissues of fish. Larger and older fish tend to have the highest levels of mercury.

CHROMIUM

Exposure to chromium occurs from ingesting contaminated food or drinking water or breathing contaminated workplace air. Chromium (VI) at high levels can damage the nose and can cause cancer

When Chromium Enters the Environment

- Chromium enters the air, water, and soil mostly in the chromium (III) and chromium(VI) forms.

- In air, chromium compounds are present mostly as fine dust particles which eventually settle over land and water.
- Chromium can strongly attach to soil and only a small amount can dissolve in water and move deeper in the soil to underground water.
- Fish do not accumulate much chromium in their bodies from water.

CADMIUM

Exposure to cadmium happens mostly in the workplace where cadmium products are made. The general population is exposed from breathing cigarette smoke or eating cadmium contaminated foods. Cadmium damages the lungs, can cause kidney disease, and may irritate the digestive tract.

When Cadmium Enters the Environment

- Cadmium enters air from mining, industry, and burning coal and household wastes.
- Cadmium particles in air can travel long distances before falling to the ground or water.
- It enters water and soil from waste disposal and spills or leaks at hazardous waste sites.
- it binds strongly to soil particles.
- Some cadmium dissolves in water.
- It doesn't break down in the environment, but can change forms.
- Fish, plants, and animals take up cadmium from the environment.
- Cadmium stays in the body a very long time and can build up from many years of exposure to low levels.

ARSENIC

Exposure to higher than average levels of arsenic occur mostly in the workplace, near hazardous waste sites, or in areas with high natural levels. At high levels, inorganic arsenic can cause death. Exposure to lower levels for a long time can cause a discolouration of the skin and the appearance of small corns or warts.

When Arsenic Enters the Environment

- Arsenic occurs naturally in soil and minerals and it therefore may enter the air, water, and land from wind-blown dust and may get into water from runoff and leaching.
- Arsenic cannot be destroyed in the environment. It can only change its form.
- Rain and snow remove arsenic dust particles from the air.
- Many common arsenic compounds can dissolve in water. Most of the arsenic in water will ultimately end up in soil or sediment.

- Fish and shellfish can accumulate arsenic; most of this arsenic is in an organic form called arsenobetaine that is much less harmful.

LEAD

Lead is a heavy, low melting, bluish-gray metal that occurs naturally in the Earth's crust. However, it is rarely found naturally as a metal. It is usually found combined with two or more other elements to form lead compounds

When lead enters the environment?

Lead is a naturally occurring element that people have used almost since the beginning of civilization. Human activities have spread lead widely throughout the environment-the air, water, soil, plants, animals, and man-made constructions.

Because lead is spread so widely throughout the environment, it can now be found in everyone's bodies; most people have lead levels that are orders of magnitude greater than that of ancient times and within an order of magnitude of levels that have resulted in adverse health effects.

CARCINOGENIC CHEMICALS

According to a recent study, published in a scientific journal, burning of incense in temples and homes releases Polycyclic Aromatic Hydrocarbons (PAHs), which are carcinogenic chemicals. If homes and temples are poorly ventilated, as generally is the case, the level of these toxic chemicals at times can be even more than that at a busy traffic intersection.

The study conducted by Ta Chang Ling of National Cheng Kung University of Taiwan points out that incense is often mixed with carcinogenic chemicals that cause cellular changes and which in turn can lead to cancer.

Ling says that the risk of cancer in poorly ventilated houses and temples is 40 times higher.

Besides even very low levels of certain other chemicals in incense, fragrance and scented products can lead to a wide range of deleterious effects on the human health. We use a variety of scented products like soap, deodorants, hair sprays, lotions, cosmetics, after-shave lotions and fragrances everyday. The effect of chemicals in these and other scented products becomes more pronounced when we spend much of our time indoors.

Perfumes and scented products are composed of toxic chemicals like acetone, toluene, ethyl acetate, formaldehyde and ethanol that are found on any list of toxic and hazardous chemicals. All these and other chemicals used in the manufacture of personal care products have been shown in animal studies to cause serious health problems and even death.

Many of these chemicals are also respiratory irritants and can trigger asthma, allergies, migraines, eye problems, nausea and fatigue. These chemicals can also affect people who do not have any pre-existing health problems. On many occasions it so happens that we may not have used any of these products personally but we may suffer discomfort due to our co-workers, colleagues or family members having used such a product.

Much work has been done on passive smoking but no attention is being paid to the fact that what is fragrance and perfume to one man could be injurious to the other person. All fragrance chemicals are volatile and quickly become airborne once released from their receptacles. Besides being volatile these chemicals are also affected by heat, light and moisture.

Once in the air these chemicals, due to the presence of external agents, break down into several other compounds that could be more irritating than the original compounds.

At one time it was thought that the chemicals used in cosmetics and perfumes affect only the skin. However, now it is known that the effects are not limited to skin alone. From the skin these chemicals are absorbed into the blood stream and from there they move on to other organs. Several of these chemicals are known to have got into the human food chains.

Man has used perfumes and fragrances since long, but such problems never cropped up as these products were made from natural ingredients like flowers and other plant parts. But today the use of these vanity products has increased tremendously and most of these products are manufactured using chemicals that are harmful to human health and environment.

Around 4000 chemicals are used in our day-today cosmetics and other personal products and 95 percent of these trace their origin to petroleum. More often than not manufacturers of these products also do not reveal the actual components of their products under the guise of 'trade secret'. This makes it difficult to study the adverse impacts of these chemicals on man.

All of us are exposed to a variety of deleterious and harmful chemicals in our environment, so why increase the chances of falling prey to them by using these perfumes, fragrances and so called "personal products".

Most of these products do not in any way either improve the quality of air or contribute to our hygiene. They rather add more pollutants to the air. And mind you that higher price of a particular product is no guarantee of its being user friendly.

We can protect ourselves, others in our vicinity and above all our very fragile environment, which is already under stress, by avoiding these products. We can make a beginning by observing these simple and easy to follow rules.

- Stop using all scented products unless you are sure that they contain only biological products.
- Do not have your clothes dry-cleaned.
- Do not use air-fresheners at home or in your car. Try ventilating them in natural air.
- Do not burn plastics, cardboard, magazines and newspapers.

- Keep your homes, offices clean and well ventilated; it will solve most of the problems.
- Grow more plants in and around your homes.

CHEMICAL SUBSTANCE IN ENVIRONMENT

A chemical substance is a material with a specific chemical composition. It is a concept that became firmly established in the late eighteenth century after work by the chemist Joseph Proust on the composition of some pure chemical compounds such as basic copper carbonate.

He deduced that, "All samples of a compound have the same composition; that is, all samples have the same proportions, by mass, of the elements present in the compound." This is now known as the law of constant composition.

Later with the advancement of methods for chemical synthesis particularly in the realm of organic chemistry; the discovery of many more chemical elements and new techniques in the realm of analytical chemistry used for isolation and purification of elements and compounds from chemicals that led to the establishment of modern chemistry, the concept was defined as is found in most chemistry textbooks. However, there are some controversies regarding this definition mainly because the large number of chemical substances reported in chemistry literature need to be indexed. A common example of a chemical substance is pure water; it has the same properties and the same ratio of hydrogen to oxygen whether it is isolated from a river or made in a laboratory. Some typical chemical substances are diamond, gold, salt (sodium chloride) and sugar (sucrose). Generally, chemical substances exist as a solid, liquid, or gas, and may change between these phases of matter with changes in temperature or pressure. Chemical reactions convert one chemical substance into another. Forms of energy, such as light and heat, are not considered to be matter, and thus they are not "substances" in this regard. Chemical substances (also sometimes referred to as a pure substances) are often defined as "any material with a definite chemical composition" in most introductory general chemistry textbooks. According to this definition a chemical substance can either be a pure chemical element or a pure chemical compound. However, there are exceptions to this definition; a pure substance can also be defined as a form of matter that has both definite composition and distinct properties.

The chemical substance index published by CAS also includes several alloys of uncertain composition. Non-stoichiometric compounds are a special case (in inorganic chemistry) that violates the law of constant composition, and for them, it is sometimes difficult to draw the line between a mixture and a compound, as in the case of palladium hydride.

CHEMICAL ELEMENTS

An element is a chemical substance that is made up of a particular kind of atoms and hence cannot be broken down or transformed by a chemical reaction into a different element, though it can be transmutated into another element through a nuclear reaction. This is so, because all of the atoms in a sample of an element have the same number of protons, though they may be different isotopes, with differing numbers of neutrons.

There are about 120 known elements, about 80 of which are stable - that is, they do not change by radioactive decay into other elements. However, the number of chemical substances that are elements can be more than 120, because some elements can occur as more than a single chemical substance (allotropes). For instance, oxygen exists as both diatomic oxygen (O_2) and ozone (O_3).

The majority of elements are classified as metals. These are elements with a characteristic lustre such as iron, copper, and gold. Metals typically conduct electricity and heat well, and they are malleable and ductile. Around a dozen elements, such as carbon, nitrogen, and oxygen, are classified as non-metals.

Non-metals lack the metallic properties described above, they also have a high electronegativity and a tendency to form negative ions. Certain elements such as silicon sometimes resemble metals and sometimes resemble non-metals, and are known as metalloids.

CHEMICAL COMPOUNDS

A pure chemical compound is a chemical substance that is composed of a particular set of molecules or ions. Two or more elements combined into one substance, through a chemical reaction, form what is called a chemical compound. All compounds are substances, but not all substances are compounds.

A chemical compound can be either atoms bonded together in molecules or crystals in which atoms, molecules or ions form a crystalline lattice. Compounds based primarily on carbon and hydrogen atoms are called organic compounds, and all others are called inorganic compounds. Compounds containing bonds between carbon and a metal are called organometallic compounds. Compounds in which components share electrons are known as covalent compounds. Compounds consisting of oppositely charged ions are known as ionic compounds, or salts.

In organic chemistry, there can be more than one chemical compound with the same composition and molecular weight. Generally, these are called isomers. Isomers usually have substantially different chemical properties, may be isolated and do not spontaneously convert to each other. A common example is glucose vs. fructose.

The former is an aldehyde, the latter is a ketone. Their interconversion requires either enzymatic or acid-base catalysis. However, there are also tautomers, where isomerization occurs spontaneously, such that a pure substance cannot be isolated into its tautomers. A common example is glucose, which has open-chain and ring forms. One cannot manufacture pure open-chain glucose because glucose spontaneously cyclizes to the hemiacetal form.

SUBSTANCES VERSUS MIXTURES

All matter consists of various elements and chemical compounds, but these are often intimately mixed together. Mixtures contain more than one chemical substance, and they do not have a fixed composition. In principle, they can be separated into the component substances by purely mechanical processes. Butter, soil and wood are common examples of mixtures.

Grey iron metal and yellow sulfur are both chemical elements, and they can be mixed together in any ratio to form a yellow-grey mixture. No chemical process occurs, and the material can be identified as a mixture by the fact that the sulfur and the iron can be separated by a mechanical process, such as using a magnet to attract the iron away from the sulfur.

In contrast, if iron and sulfur are heated together in a certain ratio (56 grams (1 mol) of iron to 32 grams (1 mol) of sulfur), a chemical reaction takes place and a new substance is formed, the compound iron(II) sulfide, with chemical formula FeS. The resulting compound has all the properties of a chemical substance and is not a mixture.

Iron(II) sulfide has its own distinct properties such as melting point and solubility, and the two elements cannot be separated using normal mechanical processes; a magnet will be unable to recover the iron, since there is no metallic iron present in the compound.

Chemicals Versus Chemical Substances

While the term *chemical substance* is a somewhat technical term used most often by professional chemists, the word *chemical* is more widely used in the pharmaceutical industry, government and society in general.

Thus the word *chemical* includes a much wider class of substances that includes many mixtures of chemical substances that often find application in many vocations; and is most commonly used only for artificial or processed substances, such as the products of the chemical industry.

Naming and Indexing

Every chemical substance has one or more systematic names, usually named according to the IUPAC rules for naming. An alternative system is used by the Chemical Abstracts Service (CAS). Many compounds are also known by their more common, simpler names, many of which predate the systematic name. For example, the long-known sugar glucose is now systematically named 6-(hydroxymethyl) oxane-2,3,4,5-tetrol. Natural products and pharmaceuticals are also given simpler names, for example the mild pain-killer Naproxen is the more common name for the chemical compound (S)-6-methoxy-α-methyl-2-naphthaleneacetic acid.

Chemists frequently refer to chemical compounds using chemical formulae or molecular structure of the compound. There has been a phenomenal growth in the number of chemical compounds being synthesized (or isolated), and then reported in the scientific literature by professional chemists around the world. An enormous number of chemical compounds are possible through the chemical combination of the known chemical elements.

At the last count, about thirty million chemical compounds are known. The names of many of these compounds are often nontrivial and hence not very easy to remember or cite accurately. Also it is difficult to keep the track of them in the literature. Several international organizations like IUPAC and CAS have initiated steps to make such tasks easier.

CAS provides the abstracting services of the chemical literature, and provides a numerical identifier, known as CAS registry number to each chemical substance that has been reported in the chemical literature (such as chemistry journals and patents).

Table. Identification of a Typical Chemical Substance

Common name	Systematic name	Chemical formula	Chemical structure	CAS registry number	In ChI
Alcohol, or Ethyl alcohol	Ethanol	C_2H_5OH		[64–17–5]	$1/C_2H_6O/c1–2–3/h_3H,^2H_2,^1H_3$

COMMON HAZARDOUS AND TOXIC AGENTS

Many people don't realise it but there are a lot of common household items that are considered to be hazardous materials. These include medications, paint, motor oil, antifreeze, auto batteries, lawn care products, pest control products, drain cleaners, pool care products such as chlorine and acids, and household cleaners. Some household cleaners may be harmful separately or when combined such as ammonia and bleach.

The vast majority of chemical cleaners pose environmental, health and safety concerns to the air, water, animals, plants and humans. Pollution caused by chemical runoff has contaminated 40per cent of the nation's waterways remain too polluted for fishing and swimming.

HOUSEHOLD HAZARDOUS MATERIALS

A wide range of health effects are linked to materials used, depending on the substance, the dose, the duration of exposure, and the susceptibility of the person exposed.

Many solvents affect the central nervous system and are skin and eye irritants. Most are flammable; many are linked to long-term adverse health effects such as liver damage. Several are known or suspected carcinogens such as benzene and toluene.

Dusts/fibres are eye and respiratory irritants, and may aggravate asthma and provoke allergies. Specific hazards: silica in clay dust causes lung disease over years of exposure; talc (white clays) may be contaminated with asbestos, a known carcinogen; some hardwood dusts lead to nasal and sinus cancers in woodworkers. Heavy metals are hazardous both as dusts and as fumes. Lead affects the nerves, digestive system, muscles and joints. Arsenic, cadmium and chromium are known carcinogens. Mercury, copper, cobalt, silver, manganese, selenium and zinc are all acutely toxic.

Acids are corrosive to skin and eyes. Acid vapors are irritating to the lungs and inhalation of small amounts may damage lung tissue. Concentrated acids can react with many other materials. Gases generated from kilns, welding or sculpting with plastics are acutely toxic; some may lead to long-term lung damage with repeated exposure.

BASIC SAFETY RULES

Know the hazards of the materials you're working with. Read the labels, request material safety data sheets (MSDS) on new products, know what precautions, safety gear and clean up procedures are advised.

Use the safest materials and procedures possible. Stay current on the new developments in your art or craft; safer, less-toxic alternatives are being devised for many activities.

Use good ventilation at all times. Local exhaust is the best, such as a hood or spray booth that vents to the outside. Next best is to use exhaust fans that pull the contaminated air away from you and exhaust it outside (an air-conditioning system is not adequate, since it recirculates most of the air). An open window usually does not provide adequate ventilation; toxins may be blown back into your face. Use good hygiene and housekeeping. Separate work and living areas; avoid eating, drinking or smoking in the work area; don't store materials in food containers; and wash and change clothes after working. Wet mop or vacuum for cleanup of dusts.

Special precautions are needed for children's art. In general, children over the age of 12 can understand and consistently follow safety instructions for the more toxic materials; younger children cannot and should use only the safest materials.

Products Hazardous Ingredients Dangers Alternatives

- Aerosol sprays butanol, butane, propanol flammable, irritant, explosive pump-type sprays, potpourri.
- Ammonia-based cleaners ammonia, ethanol irritant, toxic, corrosive (Forms poison gas when mixed with bleach.) vinegar, salt and water for surfaces and baking soda and water for the bathroom.
- Antifreeze ethylene glycol toxic (especially to pets) unknown (use caution, take to a collection centre).
- Batteries sulfuric acid, lead corrosive, toxic unknown (use caution, recycle).

- Brake fluid glycol ethers, heavy metals flammable, toxic unknown (take to a collection centre).
- Disinfectants diethylene glycol, sodium, hypochlorite, phenols corrosive, toxic 1/2 cup borax in 1 gallon of water.
- Drain opener sodium hypochlorite, sodium/potassium hydroxide corrosive, toxic plunger, flush w/boiling water, 1/4 cup baking soda.
- Flea repellant carbamates, organophosphate, pyrethrins toxic eucalyptus leaves where pet sleeps, brewer's yeast in diet.
- Floor/furniture polishes diethylene glycol, petroleum distillates, nitrobenzene flammable, toxic 1 part lemon juice w/2 parts olive or vegetable oil.
- Furniture stripper acetone, methyl ethyl ketone, toluene xylenes flammable, toxic sandpaper.
- Latex paint resins, glycol, ethers, esters flammable lime stone based white wash or casein based paint.
- Oil based paints ethylene, aliphathydro-carbons, petroleum distillates flammable, toxic latex or water based paints.
- Oven cleaner potassium, sodium hydroxide, ammonia, lye corrosive, toxic baking soda and water, salt on spills that are still warm.
- Photographic chemicals silver, acetic acid, ferrocyanide, hydro quinone corrosive, toxic, irritant unknown (use caution, take to collection centre).
- Pool chemicals muriatic acid, sodium hypochlorite algicide corrosive, toxic unknown (use caution, use until gone take to a collection centre).
- Rat and mouse killer lead arsenate, coumarins (warfarin) strychnine toxic remove food and water sources, clear harbourage, cover holes and drains rats may enter, use mechanical traps. Get a cat.
- Roach and ant killer organo-phosphates, carbamates toxic roaches: traps, boric acid. Ants: chili pepper/cream of tartar in ants' path.
- Rug and upholstery cleaners naphthalene, oxalic acid, diethylene glycol irritant, toxic, corrosive dry corn starch sprinkled on rug then vacumed up.
- Toilet bowl cleaner muriatic or oxalic paradi-chlorobenzene calcium hypochlorite irritant, toxic, corrosive toilet brush and baking soda; mild detergent.
- Thinners and turpentine n-butyl alcohol, isobutyl keytone, petroleum distillates flammable use water with water based paints.
- Transmission fluid hydrocarbons mineral spirits flammable, toxic unknown (take to a collection centre).
- Used oil hydrocarbons, (*e.g.* benzene) heavy metals flammable, toxic unknown (recycle).

Note: It is illegal to dispose of oil on/in the ground.

CHLORINE (SODIUM HYPOCHLORITE)

In paper products, such as toilet paper and paper towels:

- Bleaching paper products with chlorine bleach causes the formation of dioxin, an extremely toxic and persistent chemical known to cause cancer and disrupt the endocrine system.
- Chlorine-free toilet paper and paper towels are available at many natural food stores. Additionally, newspaper can be used in place of paper towels for cleaning windows, and rags can be used for other surfaces.

In Cleaning Products

Many household cleaners contain chlorine bleach. Chlorine bleach, or sodium hypochlorite, is a lung and eye irritant. If mixed with ammonia or acid-based cleaners (including vinegar), chlorine bleach releases toxic chloramine gas. Short-term exposure to this gas may cause mild asthmatic symptoms or more serious respiratory problems.

- To be on the safe side, don't mix chlorine bleach with anything — or just avoid chlorine bleach altogether. The EPA recommends using a non-chlorine bleach such as hydrogen peroxide to bleach clothes.

PHOSPHATES

Phosphates are minerals that act as water softeners. Although they are very effective cleaners, phosphates also act as fertilizers. When cleaning products go down the drain, phosphates are discharged into rivers, lakes, estuaries, and oceans.

In lakes and rivers especially, phosphates cause a rapid growth of algae, resulting in pollution of the water.

Many states have banned phosphates from household laundry detergents and some other cleaning products. Automatic dishwasher detergents are usually exempt from phosphate restrictions, and most major brands contain phosphates.

Some phosphate-free alternatives are available and hand dishwashing liquids do not contain phosphates.

ALKYLPHENOLS AND THEIR DERIVATIVES

Alkylphenol Ethoxylates are found in some laundry detergents, disinfecting cleaners, all-purpose cleaners, spot removers, hair colours and other hair-care products, and spermicides.

- Alkylphenol Ethoxylates are endocrine disruptors.
- Alkylphenols are produced in the environmental breakdown of alkylphenol ethoxylate surfactants, are slow to bio-degrade and have been shown to disrupt the endocrine systems of fish, birds, and mammals.

VOLATILE ORGANIC COMPOUNDS

Organic chemicals are widely used as ingredients in household products. All of these products can release pollutants while you are using them, and, to some degree, when they are stored.

- EPA's Total Exposure Assessment Methodology (TEAM) studies found levels of about a dozen common organic pollutants to be 2 to 5 times higher inside homes than outside, regardless of whether the homes were located in rural or highly industrial areas.
- Additional TEAM studies indicate that while people are using products containing organic chemicals, they can expose themselves and others to very high pollutant levels, and elevated concentrations can persist in the air long after the activity is completed.
- Many organic compounds are known to cause cancer in animals; some are suspected of causing, or are known to cause, cancer in humans.
- Some of the hazardous volatile organic compounds (VOCs) that frequently pollute indoor air — such as toluene, styrene, xylenes, and trichloroethylene — may be emitted from aerosol products, dry-cleaned clothing, paints, varnishes, glues, art supplies, cleaners, spot removers, floor waxes and polishes and air fresheners.

High levels of toluene can put pregnant woman at risk of having babies with neurological problems, retarded growth, and developmental problems. Xylenes may also cause birth defects.

Trichloroethylene is one of the chemicals suspected of causing a cluster of childhood leukemia cases due to drinking water contamination in the town of Woburn, Massachusetts, in the early 1980s. The subsequent lawsuit against the polluting company was the subject of the 1995 book and 1998 film, A Civil Action.

Styrene is a suspected endocrine disruptor, a chemical that can interfere, block or mimic hormones in humans or animals.

- VOCs such as xylene, ketones, and aldehydes are found in many aerosol products and air fresheners. Researchers found that babies less than six months old in homes where air fresheners are used on most days had 30 percent more ear infections than those exposed less than once a week.
- Levels of formaldehyde in air as low as 0.1 ppm (0.1 part formaldehyde per million parts of air) can cause watery eyes, burning sensations in the eyes, nose and throat, stuffy nose, nausea, coughing, chest tightness, wheezing, skin rashes and allergic reactions.
- Babies frequently exposed to aerosols had a 22 percent increase in diarrhea, and pregnant women frequently exposed to these products had 25 percent more headaches and a 19 percent increase in postnatal depression compared to those less frequently exposed.

Definitions:

Corrosive: A chemical, (solid, liquid or gas), that can cause destructive damage to body tissues at the site of contact. It can cause severe burns to the skin and can "eat through" clothing, metal and other materials.

Flammable: Can be ignited at almost any temperature. Spontaneously react with oxides.

Irritant: Causes soreness or inflammation of the skin, eyes, mucous membranes or respiratory system.

Oxidizer: An unstable chemical that can spontaneously react with flammables and releases oxygen.

Toxic: May cause injury or death upon ingestion (eating/drinking), absorption (touching) or inhalation (breathing into lungs).

5

Nanoparticles in Aquatic and Terrestrial Environments

DEFINITION OF NANOPARTICLES

There is no accepted international definition of a nanoparticle, but one given in the new PAS71 document developed in the UK is: "A particle having one or more dimensions of the order of 100nm or less".

There is a note associated with this definition: "Novel properties that differentiate nanoparticles from the bulk material typically develop at a critical length scale of under 100nm".

The "novel properties" mentioned are entirely dependent on the fact that at the nano-scale, the physics of nanoparticles mean that their properties are different from the properties of the bulk material. This makes the size of particles or the scale of its features the most important attribute of nanoparticles.

DIFFERENT ABOUT A NANOPARTICLE

There is no strict dividing line between nanoparticles and non-nanoparticles. The size at which materials display different properties to the bulk material is material dependant and can certainly be claimed for many materials much larger in size than 100nm. Definitions certainly become more difficult for materials that are a very long way from being a sphere, such as carbon nanotubes for example. One of the aims for these materials is to grow them into long tubes, certainly not 'nano' in length, but as they have a diameter in the order of 3nm for a single walled tube, they have properties that distinguish them from other allotropes of carbon, and hence can be described as 'nanomaterials'.

This sort of nanomaterial has led to the extension of the idea of nanomaterials being considered as such if any one of their structural features are on a scale of less than 100nm, that cause their properties to be different from that of the bulk material.

MANUFACTURING METHODS FOR NANOPARTICLES

Many of these nanomaterials are made directly as dry powders, and it is a common myth that these powders will stay in the same state when stored. In fact, they will rapidly aggregate through a solid bridging mechanism in as little as a few seconds. Whether these aggregates are detrimental will depend entirely on the application of the nanomaterial. If the nanoparticles need to be kept separate, then they must be prepared and stored in a liquid medium designed to facilitate sufficient interparticle repulsion forces to prevent aggregation.

NANOPARTICLES PROPERTIES

In nanotechnology, a particle is defined as a small object that behaves as a whole unit in terms of its transport and properties. It is further classified according to size: in terms of diameter, fine particles cover a range between 100 and 2500 nanometres, while ultrafine particles, on the other hand, are sized between 1 and 100 nanometres.

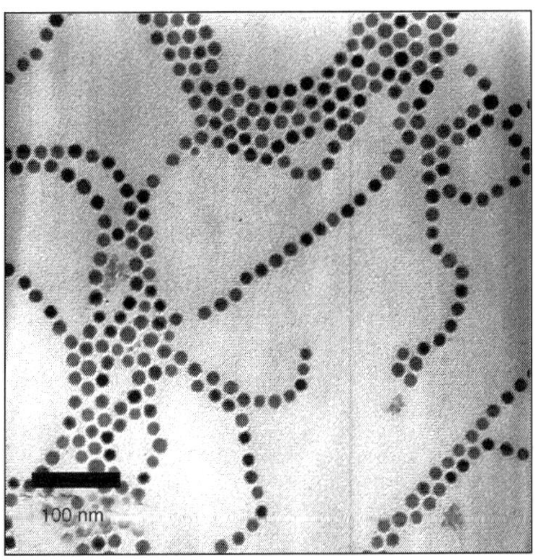

Fig. Colourized Transmission Electron Micrograph Showing Chains of Cobalt Nanoparticles.

Similar to ultrafine particles, nanoparticles are sized between 1 and 100 nanometres. Nanoparticles may or may not exhibit size-related properties that differ significantly from those observed in fine particles or bulk materials. Although the size of most molecules would fit into the above outline, individual molecules are usually not referred to as nanoparticles. Nanoclusters have at least one dimension between 1 and 10 nanometres and a narrow size distribution. Nanopowders are agglomerates of ultrafine particles, nanoparticles, or nanoclusters.

Nanometre-sized single crystals, or single-domain ultrafine particles, are often referred to as nanocrystals. Nanoparticle research is currently an area of intense scientific interest due to a wide variety of potential applications in biomedical, optical and electronic fields. The National Nanotechnology Initiative has led to generous public funding for nanoparticle research in the United States.

Although nanoparticles are generally considered an invention of modern science, they actually have a very long history. Nanoparticles were used by artisans as far back as the 9th century in Mesopotamia for generating a glittering effect on the surface of pots. Even these days, pottery from the Middle Ages and Renaissance often retain a distinct gold or copper coloured metallic glitter. This so called luster is caused by a metallic film that was applied to the transparent surface of a glazing. The luster can still be visible if the film has resisted atmospheric oxidation and other weathering.

The luster originated within the film itself, which contained silver and copper nanoparticles dispersed homogeneously in the glassy matrix of the ceramic glaze. These nanoparticles were created by the artisans by adding copper and silver salts and oxides together with vinegar, ochre and clay, on the surface of previously-glazed pottery. The object was then placed into a kiln and heated to about 600 °C in a reducing atmosphere.

In the heat the glaze would soften, causing the copper and silver ions to migrate into the outer layers of the glaze. There the reducing atmosphere reduced the ions back to metals, which then came together forming the nanoparticles that give the colour and optical effects. Luster technique showed that ancient craftsmen had a rather sophisticated empirical knowledge of materials. The technique originated in the Islamic world. As Muslims were not allowed to use gold in artistic representations, they had to find a way to create a similar effect without using real gold. The solution they found was using luster. Michael Faraday provided the first description, in scientific terms, of the optical properties of nanometre-scale metals in his classic 1857 paper. In a subsequent paper, the author (Turner) points out that: "It is well known that when thin leaves of gold or silver are mounted upon glass and heated to a temperature which is well below a red heat (~500 °C), a remarkable change of properties takes place, whereby the continuity of the metallic film is destroyed. The result is that white light is now freely transmitted, reflection is correspondingly diminished, while the electrical resistivity is enormously increased."

PHYSICAL PROPERTIES OF NANOPARTICLES

Nanoparticles are of great scientific interest as they are effectively a bridge between bulk materials and atomic or molecular structures. A bulk material should have constant physical properties regardless of its size, but at the nano-scale size-dependent properties are often observed. Thus, the properties of materials change as their size approaches the nanoscale and as the percentage of atoms at the surface of a material becomes significant.

For bulk materials larger than one micrometer (or micron), the percentage of atoms at the surface is insignificant in relation to the number of atoms in the bulk of the material. The interesting and sometimes unexpected properties of nanoparticles are therefore largely due to the large surface area of the material, which dominates the contributions made by the small bulk of the material.

For example, nanoparticles of usually yellow gold and gray silicon are red in colour; gold nanoparticles melt at much lower temperatures (~300 °C for 2.5 nm size) than the gold slabs (1064 °C); and absorption of solar radiation in photovoltaic cells is much higher in materials composed of nanoparticles than it is in thin films of continuous sheets of material – the smaller the particles, the greater the solar absorption. Other size-dependent property changes include quantum confinement in semiconductor particles, surface plasmon resonance in some metal particles and superparamagnetism in magnetic materials. Ironically, the changes in physical properties are not always desirable.

Ferroelectric materials smaller than 10 nm can switch their magnetisation direction using room temperature thermal energy, thus making them unsuitable for memory storage. Suspensions of nanoparticles are possible since the interaction of the particle surface with the solvent is strong enough to overcome density differences, which otherwise usually result in a material either sinking or floating in a liquid.

Nanoparticles also often possess unexpected optical properties as they are small enough to confine their electrons and produce quantum effects. For example gold nanoparticles appear deep red to black in solution. The high surface area to volume ratio of nanoparticles provides a tremendous driving force for diffusion, especially at elevated temperatures.

Sintering can take place at lower temperatures, over shorter time scales than for larger particles. This theoretically does not affect the density of the final product, though flow difficulties and the tendency of nanoparticles to agglomerate complicates matters. Moreover, nanoparticles have been found to impart some extra properties to various day to day products. For example the presence of titanium dioxide nanoparticles imparts what we call the self-cleaning effect, and the size being nanorange, the particles can not be observed. Zinc oxide particles have been found to have superior UV blocking properties compared to its bulk substitute. This is one of the reasons why it is often used in the preparation of sunscreen lotions., and is completely photostable.

Clay nanoparticles when incorporated into polymer matrices increase reinforcement, leading to stronger plastics, verifiable by a higher glass transition temperature and other mechanical property tests. These nanoparticles are hard, and impart their properties to the polymer (plastic). Nanoparticles have also been attached to textile fibres in order to create smart and functional clothing.

Metal, dielectric, and semiconductor nanoparticles have been formed, as well as hybrid structures (*e.g.*, core-shell nanoparticles). Nanoparticles made of semiconducting material may also be labelled quantum dots if they are small enough (typically sub 10 nm) that quantization of electronic energy levels occurs.

Such nanoscale particles are used in biomedical applications as drug carriers or imaging agents.

Semi-solid and soft nanoparticles have been manufactured. A prototype nanoparticle of semi-solid nature is the liposome. Various types of liposome nanoparticles are currently used clinically as delivery systems for anticancer drugs and vaccines.

SYNTHESIS OF NANOPARTICLES

There are several methods for creating nanoparticles, including both attrition and pyrolysis. In attrition, macro or micro scale particles are ground in a ball mill, a planetary ball mill, or other size reducing mechanism. The resulting particles are air classified to recover nanoparticles. In pyrolysis, a vaporous precursor (liquid or gas) is forced through an orifice at high pressure and burned. The resulting solid (a version of soot) is air classified to recover oxide particles from by-product gases. Pyrolysis often results in aggregates and agglomerates rather than singleton primary particles.

A thermal plasma can also deliver the energy necessary to cause evaporation of small micrometer size particles. The thermal plasma temperatures are in the order of 10,000 K, so that solid powder easily evaporates. Nanoparticles are formed upon cooling while exiting the plasma region. The main types of the thermal plasma torches used to produce nanoparticles are dc plasma jet, dc arc plasma and radio frequency (RF) induction plasmas. In the arc plasma reactors, the energy necessary for evaporation and reaction is provided by an electric arc which is formed between the anode and the cathode. For example, silica sand can be vaporised with an arc plasma at atmospheric pressure. The resulting mixture of plasma gas and silica vapour can be rapidly cooled by quenching with oxygen, thus ensuring the quality of the fumed silica produced. In RF induction plasma torches, energy coupling to the plasma is accomplished through the electromagnetic field generated by the induction coil.

The plasma gas does not come in contact with electrodes, thus eliminating possible sources of contamination and allowing the operation of such plasma torches with a wide range of gases including inert, reducing, oxidizing and other corrosive atmospheres. The working frequency is typically between 200 kHz and 40 MHz. Laboratory units run at power levels in the order of 30–50 kW while the large scale industrial units have been tested at power levels up to 1 MW. As the residence time of the injected feed droplets in the plasma is very short it is important that the droplet sizes are small enough in order to obtain complete evaporation. The RF plasma method has been used to synthesize different nanoparticle materials, for example synthesis of various ceramic nanoparticles such as oxides, carbours/carbides and nitrides of Ti and Si. Inert-gas condensation is frequently used to make nanoparticles from metals with low melting points. The metal is vaporised in a vacuum chamber and then supercooled with an inert gas stream. The supercooled metal vapour condenses into nanometre-sized particles, which can be entrained in the inert gas stream and deposited on a substrate or studied in situ.

Sol-gel

The sol-gel process is a wet-chemical technique (also known as chemical solution deposition) widely used recently in the fields of materials science and ceramic engineering. Such methods are used primarily for the fabrication of materials (typically a metal oxide) starting from a chemical solution ("sol", short for solution) which acts as the precursor for an integrated network (or "gel") of either discrete particles or network polymers.

Typical precursors are metal alkoxides and metal chlorides, which undergo hydrolysis and polycondensation reactions to form either a network "elastic solid" or a colloidal suspension (or dispersion) – a system composed of discrete (often amorphous) submicrometer particles dispersed to various degrees in a host fluid. Formation of a metal oxide involves connecting the metal centres with oxo (M-O-M) or hydroxo (M-OH-M) bridges, therefore generating metal-oxo or metal-hydroxo polymers in solution. Thus, the sol evolves towards the formation of a gel-like diphasic system containing both a liquid phase and solid phase whose morphologies range from discrete particles to continuous polymer networks.

In the case of the colloid, the volume fraction of particles (or particle density) may be so low that a significant amount of fluid may need to be removed initially for the gel-like properties to be recognized. This can be accomplished in any number of ways. The most simple method is to allow time for sedimentation to occur, and then pour off the remaining liquid. Centrifugation can also be used to accelerate the process of phase separation. Removal of the remaining liquid (solvent) phase requires a drying process, which is typically accompanied by a significant amount of shrinkage and densification. The rate at which the solvent can be removed is ultimately determined by the distribution of porosity in the gel.

The ultimate microstructure of the final component will clearly be strongly influenced by changes implemented during this phase of processing. Afterwards, a thermal treatment, or firing process, is often necessary in order to favour further polycondensation and enhance mechanical properties and structural stability via final sintering, densification and grain growth. One of the distinct advantages of using this methodology as opposed to the more traditional processing techniques is that densification is often achieved at a much lower temperature. The precursor sol can be either deposited on a substrate to form a film (*e.g.* by dip-coating or spin-coating), cast into a suitable container with the desired shape (*e.g.* to obtain a monolithic ceramics, glasses, fibres, membranes, aerogels), or used to synthesize powders (*e.g.* microspheres, nanospheres). The sol-gel approach is a cheap and low-temperature technique that allows for the fine control of the product's chemical composition. Even small quantities of dopants, such as organic dyes and rare earth metals, can be introduced in the sol and end up uniformly dispersed in the final product. It can be used in ceramics processing and manufacturing as an investment casting material, or as a means of producing very thin films of metal oxides for various purposes. Sol-gel derived materials have diverse applications in optics, electronics, energy, space, (bio)sensors, medicine (*e.g.* controlled drug release) and separation (*e.g.* chromatography) technology.

NANOPARTICLE COLLOIDS

The term colloid is used primarily to describe a broad range of solid-liquid (and/or liquid-liquid) mixtures, all of which contain distinct solid (and/or liquid) particles which are dispersed to various degrees in a liquid medium. The term is specific to the size of the individual particles, which are larger than atomic dimensions but small enough to exhibit Brownian motion. If the particles are large enough, then their dynamic behaviour in any given period of time in suspension would be governed by forces of gravity and sedimentation. But if they are small enough to be colloids, then their irregular motion in suspension can be attributed to the collective bombardment of a myriad of thermally agitated molecules in the liquid suspending medium, as described originally by Albert Einstein in his dissertation.

Einstein proved the existence of water molecules by concluding that this erratic particle behaviour could adequately be described using the theory of Brownian motion, with sedimentation being a possible long-term result. This critical size range (or particle diameter) typically ranges from nanometres (10^{-9} m) to micrometers (10^{-6} m).

MORPHOLOGY OF NANOPARTICLES

Scientists have taken to naming their particles after the real world shapes that they might represent. Nanospheres, nanoreefs, nanoboxes and more have appeared in the literature. These morphologies sometimes arise spontaneously as an effect of a templating or directing agent present in the synthesis such as miscellar emulsions or anodized alumina pores, or from the innate crystallographic growth patterns of the materials themselves. Some of these morphologies may serve a purpose, such as long carbon nanotubes being used to bridge an electrical junction, or just a scientific curiosity like the stars shown at right. Amorphous particles usually adopt a spherical shape (due to their microstructural isotropy) – whereas the shape of anisotropic microcrystlaline whiskers corresponds to their particular crystal habit. At the small end of the size range, nanoparticles are often referred to as clusters. Spheres, rods, fibres, and cups are just a few of the shapes that have been grown. The study of fine particles is called micromeritics.

CHARACTERIZATION OF NANOPARTICLES

Nanoparticle characterization is necessary to establish understanding and control of nanoparticle synthesis and applications. Characterization is done by using a variety of different techniques, mainly drawn from materials science. Common techniques are electron microscopy (TEM, SEM), atomic force microscopy (AFM), dynamic light scattering (DLS), x-ray photoelectron spectroscopy (XPS), powder X-ray diffraction (XRD), Fourier transform infrared spectroscopy (FTIR), matrix-assisted laser desorption/ionization time-of-flight mass spectrometry (MALDI-TOF), ultraviolet-visible spectroscopy, dual polarisation interferometry and nuclear magnetic resonance (NMR). Whilst the theory has been known for over a century, the technology for Nanoparticle tracking analysis (NTA) allows direct tracking of the Brownian motion and this method therefore allows the sizing of individual nanoparticles in solution.

SAFETY OF NANOPARTICLES

Nanoparticles present possible dangers, both medically and environmentally. Most of these are due to the high surface to volume ratio, which can make the particles very reactive or catalytic. They are also able to pass through cell membranes in organisms, and their interactions with biological systems are relatively unknown. However, free nanoparticles in the environment quickly tend to agglomerate and thus leave the nano-regime, and nature itself presents many nanoparticles to which organisms on earth may have evolved immunity (such as salt particulates from ocean aerosols, terpenes from plants, or dust from volcanic eruptions). A more complete analysis is provided in the article on nanotechnology. According to the ''San Francisco Chronicle'', "Animal studies have shown that some nanoparticles can penetrate cells and tissues, move through the body and brain and cause biochemical damage they also have shown to cause a risk factor in men for testicular cancer. But whether cosmetics and sunscreens containing nanomaterials pose health risks remains largely unknown, pending completion of long-range studies recently begun by the FDA and other agencies." Diesel nanoparticles have been found to damage the cardiovascular system in a mouse model.

PLATINUM NANOPARTICLES

Platinum nanoparticles are usually in the form of a suspension or colloid of sub-micrometre-sized particles of platinum in a fluid, usually water. A colloid is technically defined as a stable dispersion of particles in a fluid medium (liquid or gas). Platinum nanoparticles can be made with sizes between about 2 and 20 nanometres (nm), depending on reaction conditions. Trillions of platinum nanoparticles are suspended in the brownish red or black coloured colloidal solution. Nanoparticles come in wide variety of shapes including spheres, rods, cubes and tetrahedra.

Due to the antioxidant properties of the platinum nanoparticles, they are the subject of substantial research with potential applications in a wide variety of areas, including nanotechnology, catalysis, medicine and the synthesis of novel materials with unique properties.

SYNTHESIS

Platinum nanoparticles are made by reduction of hexachloroplatinate. A solution is rapidly stirred while a reducing agent such as hydrogen gas is added, causing platinum ions to be reduced to neutral platinum atoms. As more and more of these platinum atoms form, the solution quickly becomes supersaturated and platinum starts to precipitate in the form of sub-nanometre particles. As the reaction proceeds, platinum atoms are added to these nuclei, which grow in size. If the solution is stirred vigorously enough, the particles will be fairly uniform in size and shape. To prevent the particles from aggregating, some sort of stabilizing agent or stabilizer that sticks to the nanoparticle surface is usually added, such as sodium polyacrylate. They can be functionalized with various organic ligands to create organic-inorganic hybrid materials with advanced functionality.

BIOLOGICAL EFFECTS

Research by Yusei Miyamoto at University of Tokyo, Japan, resulted in the use of platinum nanoparticles of the size 2-3 nm to increase the lifespan of the roundworm *Caenorhabditis elegans*. Nanoparticles may present possible safety issues both medically and environmentally. Most of these issues usually arise due to the high surface to volume ratio, which can make the particles of some metals very reactive or catalytic. In particular, inhaled nanoparticles can pose health risks, and may cause inflammation and disease in the lung. They are able to pass through cell membranes in organisms and their interactions with biological systems are relatively unknown.

THE GOLD NANOPARTICLE FOR NANOTECHNOLOGY

Nanopartz™ is recognized as both an innovator and a quality supplier of gold nanoparticle based products including Accurate™ spherical gold nanoparticles, gold nanorods, micron sized gold (Microgold™), gold nanowires, and the newly released constructs: Ramanprobes™ and Plasmonic Fluorophores™. Nanopartz™ also offers platinum and palladium coated gold nanoparticles. Products developed at Nanopartz™ constitute an essential core technology for the life sciences in diagnostics, imaging, and therapeutics-particularly cancer. In other fields, Nanopartz™ products are well suited for sensors, solar cells, liquid crystals, non-linear optics, polarizers, negative refractive index materials, standards, catalytics, and as robust subcomponents in electronics. Nanopartz™ offers products for the researcher as well as the materials manufacturer.

Over the past four years, Nanopartz™ nanomaterials have appeared in over 200 publications, 2 covers, and 18 patents. Numerous companies have employed Nanopartz™ innovative nanomaterials into their new technologies and improved old technologies. Partnering with Nanopartz™ not only ensures innovation, our audited ISO 9000 system ensures repeated success.

POTENTIAL ADVANTAGES OF NANOPARTICLE DRUG DELIVERY SYSTEMS

Nanoparticles are made from biocompatible and biodegradable materials such as polymers, either natural (*e.g.*, gelatin, albumin) or synthetic (*e.g.*, polylactides, polyalkylcyanoacrylates), or solid lipids. In the body, the drug loaded in nanoparticles is usually released from the matrix by diffusion, swelling, erosion, or degradation. Nanoparticles for the purpose of drug delivery are defined as submicron (< 1μm) colloidal particles. This definition includes monolithic nanoparticles (nanospheres) in which the drug is adsorbed, dissolved, or dispersed throughout the matrix and nanocapsules in which the drug is confined to an aqueous or oily core surrounded by a shell-like wall. Alternatively, the drug can be covalently attached to the surface or into the matrix. The following are among the important technological advantages of nanoparticles as drug carriers: high stability (*i.e.*, long shelf life); high carrier capacity (*i.e.*, many drug molecules can be incorporated in the particle matrix); feasibility of incorporation of both hydrophilic and hydrophobic substances; and feasibility of variable routes of administration, including oral administration and inhalation. These carriers can also be designed to enable controlled (sustained) drug release from the matrix. The methods for nanoparticle preparation and characterization were addressed in numerous reviews. The major data on nanoparticulate formulations of the anti-TB drugs.

Table. Drug Release and Therapeutic Efficacy of the Nanoparticle-Based Formulations of the First-Line Antituberculous Drugs-Rifampin, Isoniazid, and Pyrazinamide

Delivery System	*Animal Model*	*Administration Route*	*Plasma*	*Duration of Drug Release (d) Organs*	*Regimen Producing Sterilizing Effect Ref in Lungs and Spleen*
PLG nanoparticles	Mice	Oral	6–9	9–11	5 doses every 10 d 12
	Mice	Subcutaneous	32	36	Single injection 30
	Guinea pigs	Aerosol	4–9	up to 10 d (each drug)	5 doses every 10 d 17

	Guinea pigs	Oral	4–9	up to 10 d (each drug)	5 doses every 10 d 17
Lectin-functionalized PLG nanoparticles	Guinea pigs	Oral	7–13	up to 15 d (each drug)	3 doses fortnightly 17
	Guinea pigs	Aerosol	6–14	up to 15 d (each drug)	3 doses fortnightly 17
Solid lipid nanoparticles	Guinea pigs	Aerosol	5	7	7 doses weekly 22

- *Definition of Abbreviations:* PLG = Poly(lactide-co-glycolide);
- *The Drug-to-polymer Ratio is 1*:1 for each drug

ORAL ADMINISTRATION OF NANOPARTICLE-BASED TB DRUGS

Constancy of nanoparticles offers the possibility of oral administration. The fate of nanoparticles in the gastrointestinal tract has been investigated in a number of studies. In general, the uptake of nanoparticles occurs as follows: by transcytosis via M cells, by intracellular uptake and transport via the epithelial cells lining the intestinal mucosa, by uptake via Peyer's patches.

Pandey and colleagues demonstrated that the nanoparticles provided sustained release of the anti-TB drugs and considerably enhanced their efficacy after oral administration. Three frontline drugs, rifampin (RMP), isoniazid (INH), and pyrazinamide (PZA) were coencapsulated in poly(lactide-co-glycolide) (PLG) nanoparticles. After a single oral administration of this formulation to mice, the drugs could be detected in the circulation for 4 d (RMP) and 9 d (INH and PZA); therapeutic concentrations in the tissues were maintained for 9 to 11 d. In contrast, free (unbound) drugs were cleared from the plasma within 12 to 24 h after administration. Treatment of *M. tuberculosis*–infected mice with the nanoparticle-bound drugs (five oral doses every 10th day) resulted in complete bacterial clearance from the organs. Free drugs were able to produce bacterial clearance only after daily administration of 46 doses. Similar efficacy of the nanoparticle-bound drugs was also observed in guinea pigs.

At the same time, incorporation in microparticles was less effective: their drug-loading capacity was lower as well as the plasma half-life of the bound drugs. The behaviour of polymeric nanoparticles in the gastrointestinal tract is influenced by their bioadhesive properties; adhesion of nanoparticles to the mucosa enhances the absorption of the associated drug, thus increasing its bioavailability. Thus, lectins have been shown to improve mucoadhesion of the drug due to the biorecognition of the lectin-grafted carriers by glycosylated structures in the intestine.

The efficacy of PLG-based formulations of anti-TB drugs were further improved by covalent attachment of wheat germ agglutinin. Oral administration of wheat germ agglutinin–coated PLG nanoparticles loaded with RIF, INH, and PZA in mice produced considerably extended serum half-life: detectable RIF serum levels were observed for 6 to 7 d and INH and PZA for 13 to 14 d (vs. 4–6 d and 8–9 d for nonmodified nanoparticles). All three drugs were present in lungs, liver, and spleen for 15 d. The lectin-modified formulations produced bacterial clearance in these organs after three oral doses administered every 14 d (vs. 45 daily doses of free drugs).

As suggested by the authors, the prolonged circulation of drugs encapsulated in wheat germ agglutinin–grafted nanoparticles might be attributed to the fact that lectins enhance prolonged adhesion of the particles to the intestinal surface to allow an increase in the time interval available for absorption and a localized increase in the concentration gradient between luminal and serosal sides of the membrane.

POTENTIAL FOR THE INHALATION FORM FUNCTION

The potential advantages of direct delivery of the TB drug to the lungs include the possibility of reduced systemic toxicity, as well as achieving higher drug concentration at the main site of infection. Moreover, in contrast to the oral route of administration, inhaled drugs are not subjected to first-pass metabolism. A possible obstacle to using nanocarriers for pulmonary delivery is that their mass median aerodynamic diameter, an essential parameter for the particle deposition in the lungs, is often too small.

Nevertheless, the effectiveness of pulmonary drug delivery using nanoparticles was demonstrated in a number of studies. The pharmacokinetics and antibacterial effect of the nanoparticle-bound anti-TB drugs administered via respiratory route was investigated in guinea pigs. The dose was delivered via a suitable facemask connected to the compressor–nebulizer system. A single nebulization of RMP, INH, and PZA coencapsulated in PLG nanoparticles to guinea pigs resulted in sustained therapeutic drug levels in the plasma for 6 to 8 d and in the lungs for up to 11 d. This effect was similar to that obtained after oral administration of the nanoparticulate formulation of the same drugs. In nebulization of nanoparticles to *M. tuberculosis*–infected guinea pigs at every 10th day, no tubercle bacilli could be detected in the lung after only five doses of treatment, whereas 46 daily doses of orally administered drug were required to obtain an equivalent therapeutic benefit.

Administration to infected guinea pigs of nebulized RMP, INH, and PZA coencapsulated in wheat germ agglutinin–functionalized PLG nanoparticles was even more effective: three doses administered fortnightly for 45 d were sufficient to produce a sterilizing effect in lungs and spleen.

A sterilizing effect was also achieved when the drugs were loaded in solid lipid nanoparticles. No tubercle bacilli could be detected in the lungs/spleen after seven doses of treatment of infected guinea pigs with drug-loaded solid lipid nanoparticles. It is noteworthy that the solid lipid nanoparticles display important advantages, such as the composition (physiologic compounds) and the possibility of large-scale production favoured by the feasibility to avoid organic solvents in the manufacturing process.

Intravenous Organization

In contrast to microparticles with a diameter of more than 1 μm that cannot be administered via intravascular routes, nanoparticles are small enough to allow intracapillary passage followed by an efficient cellular uptake. When administered intravenously, the nanoparticles follow the route of other foreign particulates, including intracellular pathogens. They are endocytosed by resident macrophages of the mononuclear phagocyte system and by circulating monocytes. On the other hand, in the case of infections caused by intracellularly persisting microbes (*e.g.*, *Brucella, Salmonella, Listeria, Mycobacterium*), macrophages become reservoirs for pathogens, thus representing one of the targets for delivery of antimicrobial agents.

Preferential uptake of nanoparticles by macrophages (mainly by Kupffer cells in the liver) is achieved by the physicochemical properties of the carrier and by physiologic opportunity, thus representing an example of passive delivery. This technology improves drug delivery to macrophages, increasing the amount of the drug reaching this target site, which allows reduction of the overall therapeutic dose and decrease of the adverse effects. Accordingly, the enhanced efficacy of the nanoparticle-bound antibiotics was demonstrated in a number of experimental infections.

The potential of macrophage-targeting strategy in development of the nanoparticle-based TB drugs is supported by the *in vitro* data. Incorporation of INH and streptomycin in poly(butyl cyanoacrylate) nanoparticles not only increased the intracellular accumulation (or association) of these drugs in the cultivated human blood monocytes but also produced enhanced antimicrobial activity of these agents against intracellular *M. tuberculosis* compared with their activity in extracellular fluid. Similarly, the encapsulated ciprofloxacin and RMP produced the enhanced effect against mycobacteria in the infected macrophages. This is in contrast to the previous *in vitro* observations showing that enhanced intracellular accumulation of drugs in macrophages is rarely associated with a simultaneous increase of activity against intracellular mycobacteria (compared with activity against the extracellular bacterial population). This is presumably due to the fact that the drugs and bacteria are sequestered in different intracellular compartments. It is noteworthy that the enhanced cell uptake and activity against intracellular bacteria was demonstrated for the nanoparticulate streptomycin, an aminoglycoside agent, which, in the free form, has poor intracellular access.

Clofazimine, a riminophenazine compound, is an agent considered for treating patients with *M. avium* infection. However, use of this drug was restricted because of its poor solubility. A relatively new approach was applied to solve the problem: clofazimine was formulated as a nanosuspension consisting only of the drug and a minimum amount of surfactants (particle size, 385 nm). Intravenous injection of the nanocrystalline formulation of clofazimine resulted in a considerable reduction of bacterial loads in the liver, spleen, and lungs of mice infected with *M. avium*. This result correlated with the pharmacokinetic data: drug concentrations in these organs reached high concentrations, well in excess of the minimal inhibitory concentration for most *M. avium* strains. Interestingly, the effects of the nanocrystalline formulation of clofazimine were similar to those of the liposomal formulation used as a control in this study. This study is a vivid example of application of nanotechnology for overcoming the solubility problems of poorly soluble drugs.

More details on the application of nanosuspensions in drug delivery can be found in a review on this subject. Intravenous administration of the nanoparticles has the further advantage of passive drug delivery to inflammatory sites where the endothelium becomes permeable due to pathologic processes. In this case, passive delivery of the nanoparticles would be realised by pathophysiologic opportunity. The basic principles for engineering of the nanocarriers suggest that passive accumulation in the sites with leaky vasculature would be more effective with the long-circulating (stealth) nanoparticles. These particles evade resident macrophages in the liver and have higher probability to reach other sites in the body.

Other Routes of Organization

The nanoparticle-based formulations was further demonstrated by effective subcutaneous treatment of mice infected with *M. tuberculosis*. A single subcutaneous dose of PLG nanoparticles loaded with RMP, INH, and PZA maintained therapeutic drug levels in plasma for 32 d and in lungs/spleen for 36 d. Moreover, this single subcutaneous injection produced a sterilizing effect in lungs and spleen of the infected mice (36 d post treatment), thereby demonstrating a better chemotherapeutic efficacy, as compared with daily treatment using free drugs (35 oral doses). As suggested by the authors, the nanoparticles form a depot at the injection site that is slowly releasing drugs into the circulation.

POPULATIONS IN AQUATIC ENVIRONMENTS

A group of organisms occupying a given space at a particular moment in time is called a *population*. A population may consist of a single species and thus constitute a *species population*, or it may be composed of a number of species exhibiting similar ecological traits (food requirements, breeding sites, etc.) and regarded as a *mixed population*.

In addition to structure and time-space attributes, a population also exhibits a number of measurable characteristics such as birth rate (natality), death rate (mortality), density, and capacity for increase. The extent to which these traits are manifested in an environment is strongly influenced by properties of that environment.

THE PRINCIPLE OF LIMITING FACTORS

The very fact that we can recognize an environment-organism complex such as a lake community or an estuarine community attests to the premise that all organisms do not react uniformly to all combinations of physical, chemical, and biological features of the environment.

Through evolutionary processes, organisms have become variously adapted to certain sets of conditions. It follows, therefore, that successful development and maintenance of a population depend upon harmonious ecological balance between environmental conditions and tolerance of the organisms to variations in one or more of these conditions. This idea suggests that one or more factors can serve to limit the areal distribution, density, and other attributes of a population. A factor that so checks or exerts some restraining influence upon a population through incompatibility with species requirements or tolerance is said to be a *limiting factor*. You will note that the limiting factor principle rests essentially upon two basic concepts. One of these relates organism to environmental supply of materials needed for metabolism and growth. The second concept pertains to the tolerance which an organism exhibits towards environmental factors and conditions.

The aspect of environmental supply in the limiting factor principle includes, among other things, the provision of nutrients and other substances (respiratory gases, for example) necessary for growth of the organisms. If, out of a broad array of materials required for growth and development, one is absent, a given species will not be able to survive. On the other hand, if the nutrient is present but only in small amounts, the population of a particular species will be limited proportionately. These conditions were described in 1840 by Liebig in what has come to be called the *law of the minimum*. This principle recognizes that the development of a population is essentially regulated by the substance occurring in minimal quantity relative to the requirement of the population.

It is frequently difficult to determine precisely which of the factors of a complex is exerting a limiting effect. This is due largely to the fact that the utilization of a certain substance may be regulated by other materials, or by physical factors in the environment, or by physiological adjustments of the organism itself. It has been shown, for example, that the uptake of phosphorus by the alga *Nitzschia closterium* is influenced by the quantity of nitrate and phosphate in the environment; nitrate utilization, on the other hand, appears to be unaffected by the phosphate. Potassium, calcium, and magnesium generally regulate interactions in plant metabolism. The assimilation of nutrients by some algae is related to environmental temperature. Temperature and carbon dioxide tension may affect the rate of oxygen utilization in certain fishes. We have previously learned that the absence of free carbon dioxide in waters is not necessarily critical for certain plants which can assimilate carbon dioxide from bicarbonates.

Dissolved oxygen is necessary in respiration of most aquatic organisms. The minimal requirements of species vary and often limit the spatial distribution of certain forms in the community. The water issuing from the "boil" is oxygenless, but gains the gas with downstream flow. A small amount of hydrogen sulfide is present in the boil and decreases downstream; this may also be involved in the observed distribution.

The law of the minimum pertains primarily to the reaction of organisms to factors necessary for growth and metabolism. In the case of plants, these factors include the inorganic nutrients and solar energy for photosynthesis. In addition to satisfying their metabolism needs, organisms are also confronted with physical and chemical factors which essentially regulate the distribution and numbers and the extent to which the organisms are able to utilize growth factors. Such environmental features as salinity, temperature, and currents govern many populations.

The extent to which these factors limit a population depends primarily upon the *tolerance* of the organisms to a single factor, or, in some instances, to a complex of interacting factors. These ideas are the essence of the *law of tolerance* stated in 1913 by V. E. Shelford of the University of Illinois.

Tolerance to features of the environment varies widely among aquatic organisms. A population may exhibit a wide range of tolerance towards one condition, and a narrow range towards another. All stages in the life history of an organism do not necessarily show similar ranges of tolerance. The range of tolerance towards a given factor may be modified by another factor. A wide range of distribution of a species is usually the result of the wide tolerance.

In describing the tolerance of an organism, the prefix *eury*-, meaning wide, or *steno*-, meaning close or narrow, is added to a term for the particular feature. We have already become familiar with the terms *euryhaline* (wide salt tolerance) and *stenohaline* (narrow salt tolerance). To these, we might add *eurythermal* and *stenothermal*, pertaining to temperature. Let us now consider some examples of tolerance relationships, for such are of utmost importance in regulating the abundance and distribution of organisms.

Contractile vacuoles are characteristic of freshwater protozoans, but not of their marine relatives. In certain flatworms (*Gyratrix*, for example) the flamecell mechanism serving in osmoregulation and excretion is more complexly developed in freshwater forms than in estuarine ones; in marine flatworms, major parts of the system are absent. Freshwater crayfish possess long, well developed nephridial canals; in the estuarine and marine lobsters these structures are reduced and poorly developed. The fluids of marine bony fishes are hypotonic to the environment, and therefore the animals tend to lose water to the medium. In contrast to freshwater inhabitants, marine fishes actively drink water, excrete a concentrated urine, and excrete excess salts (obtained by drinking sea water) against an environmental gradient through specialized cells in the gill region. In the estuarine environment, these marine forms must function in the manner of freshwater species. The young of certain species of marine fishes inhabit estuaries and possess glomerular kidneys; with migration to higher salinity as adults, the glomeruli degenerate. Highly euryhaline fishes, particularly those adapted to extreme migrations from fresh to salt water (the eel (*Anguilla*), for example), are usually capable of performing as either marine or freshwater fish.

We have dwelt on this subject of osmoregulation because it demonstrates some of the forms of adjustment often necessary in order for organisms to tolerate certain controlling factors in the environment. In many instances, however, the limit of tolerance of a given organism may be modified by other factors. It has been shown, for example, that the shrimp (*Crangon crangon*) and estuarine crab (*Carcinus maenas*) are able to tolerate lower salinities during summer seasons of warm water. In Japanese waters, the oyster (*Ostrea gigas*) exhibits a wider range of salinity tolerance at winter temperatures than during the warm season.

The tendency for marine species to tolerate lower salinities at higher temperatures accounts, in part at least, for the generally greater number of species in estuaries of the tropical zones than in the cooler climes.

As in the case of salinity, temperature also acts as a controlling factor related to range of tolerance of species. In this sense, temperature serves to regulate growth and metabolic rates of various organisms, and often determines the time of reproduction. It is immediately apparent, therefore, that temperature is highly important in delimiting the rate of utilization of nutrients and light plants, and the tempo of food intake by animals, to satisfy metabolic demands. These relationships are contained in the "Q_{10} Rule," or Van't Hoff's principle, a chemical law applied to some physiological processes.

The principle holds that the rate at which biological processes proceed is increased nearly two-fold with each 10° rise in temperature (the rate is a linear function of temperature). Obviously, this principle operates only within the range of tolerance of a given species, and is further restricted by an optimal point within the over-all range. For example, if sufficient nutrients are available, an algal population may grow rapidly and "bloom" at the optimal temperature for that species; beyond this point, and before the maximum tolerance limit is reached, the population declines.

Throughout the living world, organisms have become adapted to temperature ranges and fluctuations in a spectacular variety of ways. In the first place, we recognize two broad divisions:

- The "warm-blooded" (*homoiothermic*) forms, the birds and mammals, whose body temperatures are maintained at a uniform level independent of environment; only a few of this group are wholly aquatic,
- The "cold-blooded" (*poikilothermic*) organisms whose body temperatures approximate that of the environment and vary accordingly. Among the latter group are such extremes as the algae that live in ice water of the polar regions and bacteria that inhabit hot springs at temperatures near 90°C. Between these extremes are found many species of plants and animals variously adapted to thermal conditions.

Great numbers of species are unable to exist as physiologically active stages throughout the range of temperatures experienced in their environment; these typically form spores, or other "resting stages," which exhibit a wider range of tolerance than do the active forms.

Stenothermal and eurythermal species are found in all major taxonomic groups. The European turbellarian, *Crenobia alpina*, is apparently restricted to a temperature regime below 12°C; sessile rotifers are usually found at temperatures above 15°C; tardigrades are active in a wide range from about 0° to 40°C; the optimum temperature for the bryozoan, *Pectinatella*, is between 20° and 23°C, the lower limit of tolerance being near 15°C; the oyster (*Crassostrea virginica*) tolerates temperatures from 4° to 34°C.

We quickly call to mind typical "cold-water" and "warm-water" fishes, many of these being restricted to such environments by the temperature tolerance of the developing eggs; the eggs of the eastern brook trout (*Salvelinus fontinalis*) generally fail to hatch at temperatures above 12°C, the optimum being near 8°C. The optimum temperature for growth of the adult trout is about 15°C. Thus, several optima may be operative in the life history of a given species.

Light appears to be a controlling factor in the distribution of certain animals in time and space. Diurnal vertical migrations of planktonic crustaceans have been noted and studied widely.

These movements characteristically take the organisms to near-surface regions during the middle of the night, and to the depths during midday. Numerous explanations have been suggested, and, indeed, the causes may be numerous.

However, the correlation between the depth of penetration of the blue segment of the light spectrum and the position of greatest abundance of a copepod species, *Cyclops strenuus*, would indicate a strong influence of the light. Note particularly the conspicuous change in space occupancy with increasing day length from April to June.

From consideration of thc "law of the minimum" as it pertains to nutrients, and the "law of tolerance" as it relates to environmental "controls" we can now appreciate the broader *principle of limiting factors*. This principle embodies the concept that population growth and success, in terms of abundance and distribution, are determined by a set of environmental factors, any one of which may, through scarcity *or overabundance*, be limiting.

We have seen a number of examples in which short supply of a requirement could hinder the success of a population; this is not difficult to understand. What is sometimes more difficult to comprehend is that an excess of normally important factors such as nutrient salts or sunlight can exert lethal effects upon organisms. For example, high concentrations of nitrate and phosphate have been shown to be toxic to both plants and animals.

Several exciting aspects of the limiting factors principle can be shown in the reactions between sunlight and phytoplankton in photosynthesis. Note first that in both cases the rate of photosynthesis (measured by rate of uptake of carbon dioxide by plants) is inhibited by "too much" as well as by "too little" light. Light conditions optimums for photosynthesis are indicated at the peak of each curve, but notice that these conditions differ for the two seasonal assemblages.

It appears also that the spring phytoplankters exhibit a wider range of tolerance to the light intensities encountered than do those of autumn. Of particular interest is the higher rate of photosynthesis in the dim autumn light than in the relatively bright light of spring. This would seemingly indicate that the autumn community utilizes light more efficiently than do the spring populations. In all probability, however, the species comprising the communities are different.

The principle of limiting factors is not built upon physical and chemical factors alone. Biological features of the environment are often quite influential in determining the abundance and distribution, or over-all success, of a specific population. Once we introduce other organisms and their demands and activities into a set of factors, we immediately perceive greatly increased complexity of ecological relationships. For in addition to reactions between a population and the physicochemical factors, we must also recognize coactions between, or among, the several species living together. We must also recognize coactions among the individuals of any given population.

Interspecific relationships involve a broad spectrum of coactions, including several forms of symbiosis, tolerance, competition for one or many of a great variety of necessities, antibiosis, parasitism, and predation. Within this list of activities are found those that pertain to shelter, food supply and procurement, and breeding activities.

These often relate plant to plant, or plant to animal, or animal to animal in a broadly graded series of coactions ranging from those in which one member of the relationship suffers extreme harm at the expense of another, to those in which both of the participants are benefited. Singly, or in combination, interspecific relations may, and often do, act as a limiting factor. They are also basic in the structure and dynamics of communities.

The ability of a species to successfully occupy a place in the economy of a community is also determined by intraspecific relations of the individuals and certain inherent biological qualities of the population. Intraspecific competition for such requirements as food, territory, and breeding partners may serve to limit a given population.

Similarly, overcrowding, which may in some cases lead to stress symptoms, can be limiting in nature. Inherent biological qualities such as frequency of breeding activity, number of offspring produced, survival rate, and mortality rate are often correlated with physiological tolerance to environmental factors. Upon these adaptations depend, at least in part, the abundance and distribution of a population. Some of these population attributes will be examined in the following paragraphs.

Although we have somewhat dissected the set of limiting factors for the sake of illustration, it is important to bear in mind that a single factor can become limiting only when it is near the maximum or minimum of the range of tolerance of a population. In nature, under usual conditions, a given population reacts to a complex of factors. The composition of this complex is not normally fixed; some of the components change seasonally, and a number of interdependencies exist, as we have seen.

A specific organism or population, on the other hand, presents a different story; its tolerance to each factor is more or less firmly established by inheritance and evolution. Thus the position and place of a species in the economy of a community is fairly well defined and delimited by relations between organisms and environmental factors. This thought brings us to another important concept in population ecology, that of the *niche*.

HUMAN NATURE ENVIRONMENTAL PHILOSOPHIES

Throughout the history of the human species, environmental philosophy has played a major role. That role has, however, undergone profound alteration, especially in Western civilization as it is manifesting itself and merging with others here in the United States.

We have gone from huntergatherer to tiller of the soil to industrialist to scientist to the age of information. Both our niche and our intellectual relationship to our ecosystems have altered accordingly. Emerging from the status of one more primate species in the wild, we have learned to fear the natural world. It is in our nature to seek conquest of whatever we fear.

Some may believe human populations have conquered the natural world, but most of us realise that our place as conqueror is tenuous at best. In recent years, we have come to realise that we may represent a unique hazard not only to our ecosystems but to the biosphere as a whole. We are rediscovering our niche and the habitats in which it is active. When humans left the forests and fields to enter into agricultural and later societies, they came to fear the wilderness. When North America was invaded by representatives of Western civilization, the frank intent was to conquer those usurping the land—wild beasts, howling wilderness, and hellish fiends. This vast wilderness was a place of evil as well as one of physical danger. Conquest was a moral imperative.

The contemporary environmental movement in the United States traces its most recent roots to the conservation and preservation movements in the early twentieth century. In fact, early conservation helped shape the notion of "commons" as a public domain in resources that are held not susceptible to private appropriation. Conservation seeks to ensure the preservation of a quality environment in company with a continuous yield of useful plant, animal, and natural materials—natural resources. Conservation, then, tends to be more anthropocentric than preservation, which tends more towards biocentrism or ecocentrism.

Our literature and the popular arts in the United States are rich in images identifiable as personal ideas of wilderness, conservation, and preservation. Among the best known of the writers is Aldo Leopold, who spoke of conservation as a "state of harmony between men and land"—his "land ethic," an ecological conscience. Ecological health, according to Leopold, lies in the land's capacity for self-renewal. Conservation arises from efforts to understand that capacity in order to preserve it. The land ethic is a major concept incorporated into the contemporary environmental movement.

THE CONTEMPORARY ENVIRONMENTAL MOVEMENT

What Is It? In the early 1960s, the environmental movement in the United States still took the form of conservation or preservation. In the 1970s, the shock of gross pollution and disappearing landscapes moved us to a kind of grim pleasure expressed in part as cultural and individual ambivalence towards progress. The environmental movement provided one outlet for our malaise. Whatever our individual relationship to the movement, we can all probably agree that the role of culture is not to overcome nature but to be open to its potential. A primary lesson from ecology is that all "poverty" is attributable to unrestrained population growth. The environmental movement, then, is seeking ways to alleviate the results of population growth, unrestrained at least at the local level, and thus to relieve ecological poverty in whatever biological systems in which it is being expressed.

An Ecological Perspective

To the extent Odum can be cited for the ecological perspective, he has observed that ecologically sound conservation, one of the missions of contemporary environmental policy, is neither a matter of hoarding resources nor one of total rejection of development. Instead, the objective should be one of rejection of unplanned development in violation of ecological laws. In the past, such ecologically unsound development has rightfully been attributed to ignorance as well as to shortsightedness and advancement of parochial interests.

Today, there is no excuse for development characterized by any of those three factors. In the United States, at least, there is a very real movement towards integration of all the factors supporting development that is ecologically responsible. It is a youthful movement, however, and subject to all the excesses of youth. At least we are moving away from all three impediments to environmental quality.

One of the major problems, and one that may be insurmountable, is the nearly total focus on the human perspective. We tend to speak in terms of environmental quality in association with products, recreation, and aesthetics. We seek to ensure continuous yields of useful plants, animals, and materials. Not until we act on the knowledge that ultimately it is only healthy environments that are healthful are we likely to change our ways in an ecologically sound manner.

As environmental professionals seek to serve what is sometimes called stewardship of ecosystems and of the biosphere itself, some turn to cultures not ordinarily included in that of mainstream America. For example, some assert that we should heed the injunction of certain Native Americans to consider the effects of our actions through seven generations. All too often, we have barely thought beyond the current generation. Another lesson to take from Native American peoples is the notion that we cannot preserve ecological systems by the sole expedient of dictating by law. There instead must be a profound alteration in the way each of us relates to the world around us.

CHALCOGENIDE NANOPARTICLES

There is a great deal of interest in the synthesis of transition-metal oxide nanoparticles because of their potential applications in catalysis, high-performance ceramics, magnetic storage, solar energy conversion, and ferrofluids. Superparamagnetic transition-metal oxide nanoparticles, usually comprising a single magnetic domain, are particularly interesting in their many biomedical applications, which include magnetic cell sorting, drug targeting experiments, and gene transfection, and in hyperthermic cancer therapy.

By the same token, the optical properties of semiconductor chalcogenide nanoparticles are also a fascinating area of study, with rapidly emerging applications as luminescent probes in biology, and in polymer-quantum dot light-emitting diode (LED) devices, to name a few. In both classes of materials, *viz.* oxide and chalcogenide nanoparticles, the size of the particles has direct bearing on their properties. In preparative techniques, therefore, the ability to control size is crucial. A few rules of thumb come into play in the preparation of nanoparticles. The first is to avoid excessively high temperatures because these would lead to sintering. The second is to cap particles at the desired size to prevent further growth and to provide stability through the coordinative saturation of dangling bonds on the particle surface. In addition, the separation of the nucleation and growth in the preparation is advisable as this results in narrower size distributions.

The use of a suitable capping agent such as a long-chain amine, a long-chain thiol, trioctylphosphine oxide (TOPO), etc. has the added advantage that these can make the particles soluble in different solvents. Solubility enables size-selective separation and the formation of superlattice assemblies.

The direct oxidation of metals is usually exothermic and results in sintering. As a result, it is usually contraindicated for the preparation of oxides in nanoparticulate form. While many oxide materials can be prepared by hydrolysis in aqueous solution, it is difficult to incorporate a long-chain organic capping agent under these conditions.

The many routes for preparation of oxide nanoparticles that have emerged recently are sensitive to these issues. Chemseddine and Moritz have demonstrated the synthesis of monodisperse anatase TiO_2 (anatase) nanoparticles of different sizes and crystal habits, by decomposing titanium alkoxides. Colvin and coworkers have prepared soluble TOPO-capped anatase TiO_2 nanoparticles, using the metathetic reaction between titanium alkoxide and titanium halide. Another interesting nonhydrolytic single precursor approach involves decomposition of metal cupferron complexes to the corresponding metal oxide in trioctylamine at 550 K.

Cupferron is *N*-nitroso *N*-phenyl hydroxylamine, and it can be complexed with a number of transition-metal ions, making this an important general route to oxide nanoparticles. In the same vein, many synthetic approaches are available for the preparation of transition-metal chalcogenide nanoparticles.

Bawendi and coworkers have demonstrated the preparation of monodisperse CdE (E = S, Se, Te) nanoparticles of different sizes using organometallic precursors and TOPO as solvent and capping agent. Green and O'Brien have reviewed the use of single source precursor to prepare quantum dot nanoparticles. The use of ethylene diamine as a solvent is an effective route to several binary and ternary chalcogenides of different sizes, ranging from the submicron to the nano.

Despite this seeming abundance of methods for the preparation of nanoparticles, the challenge remains for the synthetic chemist to develop new techniques that are environment friendly and economically viable.

One of the common features of the many synthetic techniques is the use of expensive and often toxic solvents such as TOPO or trioctylamine. It has occurred to us that ideally one should be able to use a simple and inexpensive solvent such as toluene.

The limitation with toluene is that it is low-boiling, and cannot be used under reflux for the preparation of crystalline particles as the temperature would be too low. We overcome this problem of the low boiling point of toluene by using it in a sealed autoclave (bomb) under the autogenous pressure developed on heating.

By this simple method, we are able to work with toluene as the solvent at temperatures as high as 573 K. This provides a very appropriate alternative to solvents such as TOPO. In this article, we review our recent solvothermal synthesis of amine-capped transition-metal oxide nanoparticles using cupferron complexes as precursors. We have been able to replace the expensive and toxic amine solvent used by Alivisatos and coworkers by toluene. The preparation of capped sub-10 nm γ-Fe_2O_3 and CoFe2O4 particles by solvothermal methods, and their magnetic study has been reported by us previously. Soluble 3-nm thiol-capped CdSe nanoparticles have been prepared from cadmium stearate under similar solvothermal conditions.

This work has been reported by us elsewhere. The use of stearates instead of toxic and flammable organometallic precursors is worthy of note. We also report for the first time, the solvothermal preparation of lead selenide PbSe and lead iodide PbI_2 phases, prepared from the stearates in toluene, with H_2Se or HI being generated in situ by the reduction of Se or I2 by tetralin. In this, we follow some old work by Mitchell and Morgan involving the preparation of binary chalcogenides in hot nujol using the aromatization of sitosterol by Se to generate H_2Se in situ.

OXIDE NANOPARTICLES

The preparation of amine-capped γ-Fe_2O_3, $CoFe_2O_4$, and $ZnFe_2O_4$ was achieved by the solvothermal decomposition of the corresponding metal cupferron complexes in the presence of *n*-octylamine (or *n*-dodecylamine) at elevated temperature in a teflon gasketted SS316 stainless steel autoclave.

Toluene was used as solvent with 70 per cent filling. $ZnFe_2O_4$ nanoparticles could, for example, be prepared by taking 0.6 g (1.8 mmol) of Zn-cupferron complex, 1.7 g (3.6 mmol) Fe-cupferron complex, and 6.1 cm3 of dodecylamine in 48 cm3 of toluene. The autoclave was placed in an air oven that was heated to 503 K. The reaction was carried out for 1 h, after which the autoclave was removed and allowed to cool to the room temperature. The resulting solution was brown and viscous with some suspended particles, which could be precipitated by increasing the solvent polarity (adding a polar solvent such as 2-propanol). The collection could be facilitated by placing the material taken in a beaker on a magnet. The particles were washed with 2-propanol and diethyl ether.

The other oxide nanoparticles were prepared in a similar fashion. Figure 1 displays a transmission electron micrograph of *n*-dodecylamine-capped $ZnFe_2O_4$ nanoparticles prepared by solvothermal decomposition of metal cupferron complexes. The mean diameter of the nanoparticles is 7.4 nm with a standard deviation of 0.9 nm. This is in keeping with the Scherrer broadening of the powder X-ray diffraction (XRD) patterns. The correspondence in the particle sizes obtained from the two techniques suggests that the nanoparticles are nearly single-crystalline. Rietveld refinements were performed on the powder XRD patterns using the cubic spinel structure as a model. The Rietveld fits confirm the assignment of crystal structure. The refined cell parameter was $a = 8.50(1)$ Å.

The stoichiometry of the spinel oxide could also be established by X-ray photoelectron spectroscopy and energy-dispersive analysis of X-rays (SEM/EDAX). The presence of capping agent could confirmed by using Fourier transform infrared (FTIR) spectroscopy and thermogravimetric analysis in air. Transition-metal oxide nanoparticles are often magnetic, being derived from bulk ferromagnetic spinels (in the case of $CoFe_2O_4$ or $ZnFe_2O_4$).

Because of their small size, they usually cannot support more than a single magnetic domain and hence display the phenomenon of superparamagnetism. One of the characteristics of a superparamagnet is that the magnetization (apart from showing no hysteresis) shows a scaling behaviour. In the superparamagnetic regime, plots of M vs. H/T collapse on to a single trace. In the upper panel (a) are traces of the magnetization as a function of the scaled field (H/T) at temperatures of 150, 200, 250, and 300 K. These correspond to a superparamagnetic regime, and the traces are collapsed. In the lower panel (b), M vs. H/T is displayed at the temperatures of 100 and 50 K. This is now the blocked regime (where domains have made an appearance), and the traces display hysteresis.

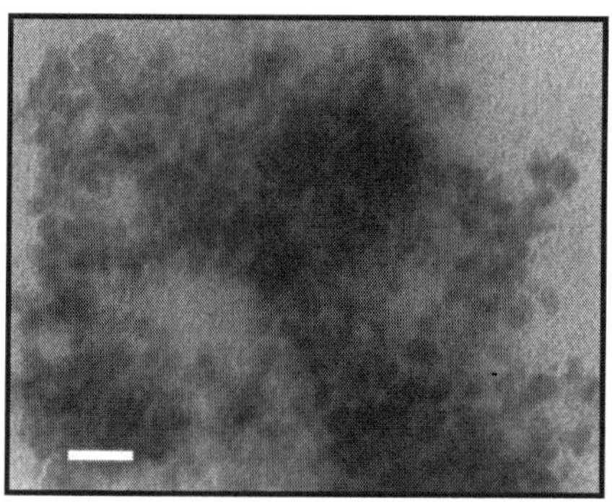

Fig. TEM Image of *n*-dodecylamine-capped 7-nm Zn Fe_2O_4 Nanoparticles. The Bar Corresponds to 20 nm.

CHALCOGENIDES AND HALIDES

For the synthesis of chalcogenide phases, the starting materials of choice were the metal stearates. These are soluble in hot toluene, and in comparison with other metal sources (such as dimethyl cadmium) are inexpensive and relatively safe. In a typical preparation of metal chalcogenide or halide, the metal stearate is taken in toluene with selenium or iodine and a small amount of tetralin. This is sealed in a stainless steel autoclave and then placed in a preheated air oven held at the required temperature. After the reaction is over, the autoclave is taken out and allowed to cool to room temperature, following which the product is collected. Cadmium stearate was prepared by reacting Cd(OH)2 with molten stearic acid at 363 K. The product was recrystallized from toluene.

For the preparation of CdSe nanoparticles, 1.36 g (2 mmol) of [Cd(St)2], 0.158 g of selenium (2 mmol), 0.2 g of tetralin (1.5 mmol), and 0.1 g (0.5 mmol) of dodecanethiol as capping agent was taken in 50 cm3 of toluene (which is 70 per cent filling in the autoclave). The autoclave was then placed in an oven preheated to 523 K where it was held for 5 h. The dark red solution obtained from the reaction precipitated CdSe nanoparticles when the solvent polarity was increased by adding the polar nonsolvent 2-propanol.

Lead stearate was prepared by adding hot aqueous solution of the metal salt to hot sodium stearate solution (prepared from stearic acid and sodium hydroxide solution). Lead stearate was immediately precipitated. After washing with copious quantities of hot water and, subsequently, with ether to remove excess stearic acid, it was characterized by XRD and IR spectroscopy.

The transmission electron microscopy (TEM) image of the CdSe nanoparticles. The inset is a histogram of the particle sizes. The mean diameter of the particles was found to be 3.0 nm with a standard deviation of 0.2 nm, which is about 6 per cent of the mean. Powder XRD patterns indicate that the nanoparticles are crystallized in the cubic zinc blend structure rather than the usual hexagonal wurtzite structure.

Rietveld refinement yields an *a* lattice parameter of 6.01(1) Å. The cubic structure is further supported by simulating the powder XRD pattern using the DIFFaX programme, which allows hcp and fcc stackings of 3×3 nm of CdSe to be considered.

Optical properties of the CdSe nanoparticles are consistent with what would be expected of particles in this size range. FTIR spectra as well as thermogravimetric analysis in air confirm the presence of capping agents. In the EDX analysis, the Cd:Se atomic ratio was found to be 1:1, while that for S:Se:Cd was found to be 1:2:2 (within a 5 per cent error). PbSe prepared solvothermally as indicated above results in rather crystalline particles with sizes in the sub-1 ìm range, rather than the nanometer range. The powder XRD pattern of these particles. The vertical lines are the expected peak positions for the cubic NaCl structure of PbSe.

PbI_2 is a layered compound, and the solvothermal preparation results in some turbostratic disorder, evidence for which are the sawtooth profiles observed for some of the reflections in the powder XRD pattern. This phase cannot be described as nano either. Both for PbSe and PbI2, it was not possible to control the growth of the particles through the use of capping agents.

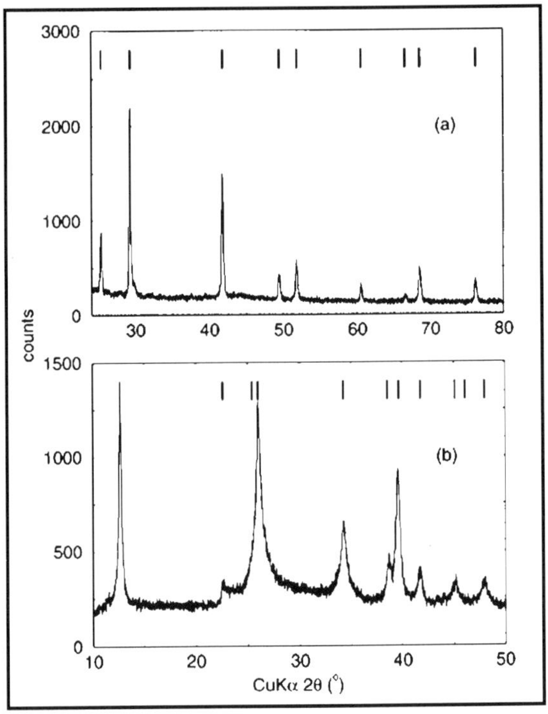

Fig. Powder XRD Patterns of Solvothermally Prepared (a) PbSe and (b) PbI2. The Vertical Lines within the Plots are the Expected Peak Positions for these Phases

6

Nanopollution

Nanopollution is a generic name for all waste generated by nanodevices or during the nanomaterials manufacturing process. This kind of waste may be very dangerous because of its size. It can float in the air and might easily penetrate animal and plant cells causing unknown effects. Most human-made nanoparticles do not appear in nature, so living organisms may not have appropriate means to deal with nanowaste. It is probably one great challenge to nanotechnology: how to deal with its nanopollutants and nanowaste.

Environmental assessment is justified as nanoparticles present novel (new) environmental impacts. Scrinis raises concerns about nano-pollution, and argues that it is not currently possible to "precisely predict or control the ecological impacts of the release of these nano-products into the environment."

Ecotoxicological impacts of nanoparticles and the potential for bioaccumulation in plants and microorganisms remain under-researched. The capacity for nanoparticles to function as a transport mechanism also raises concern about the transport of heavy metals and other environmental contaminants. A May 2007 Report to the UK Department for Environment, Food and Rural Affairs noted concerns about the toxicological impacts of nanoparticles in relation to both hazard and exposure. The report recommended comprehensive toxicological testing and independent performance tests of fuel additives. Not enough data exists to know for sure if nanoparticles could have undesirable effects on the environment. Two areas are relevant here:

- In free form nanoparticles can be released in the air or water during production (or production accidents) or as waste by-product of production, and ultimately accumulate in the soil, water or plant life.
- In fixed form, where they are part of a manufactured substance or product, they will ultimately have to be recycled or disposed of as waste.

It is not known yet whether certain nanoparticles will constitute a completely new class of non-biodegradable pollutant. In case they do, it is not known how such pollutants could be removed from air or water because most traditional filters are not suitable for such tasks (their pores are too big to catch nanoparticles). Of the US$710 million spent in 2002 by the U.S. government on nanotechnology research, only $500,000 was spent on environmental impact assessments. Risks identified by Uskokovic (2007) include: self-replicating nanobots aggressively or through slowly rising supremacy wiping out the whole biosphere; further destabilising the already endangered diversity of the biosphere.

Concerns have been raised about Silver Nano technology used by Samsung in a range of appliances such as washing machines and air purifiers.

Life Cycle Responsibility

To properly assess the health hazards of engineered nanoparticles the whole life cycle of these particles needs to be evaluated, including their fabrication, storage and distribution, application and potential abuse, and disposal. The impact on humans or the environment may vary at different stages of the life cycle.

The Royal Society report identified a risk of nanoparticles or nanotubes being released during disposal, destruction and recycling, and recommended that "manufacturers of products that fall under extended producer responsibility regimes such as end-of-life regulations publish procedures outlining how these materials will be managed to minimize possible human and environmental exposure".

Reflecting the challenges for ensuring responsible life cycle regulation, the Institute for Food and Agricultural Standards has proposed standards for nanotechnology research and development should be integrated across consumer, worker and environmental standards. They also propose that NGOs and other citizen groups play a meaningful role in the development of these standards.

FORMS OF POLLUTION

The major forms of pollution are listed below along with the particular pollutants relevant to each of them:

- Air pollution, the release of chemicals and particulates into the atmosphere. Common gaseous air pollutants include carbon monoxide, sulfur dioxide, chlorofluorocarbons (CFCs) and nitrogen oxides produced by industry and motor vehicles. Photochemical ozone and smog are created as nitrogen oxides and hydrocarbons react to sunlight. Particulate matter, or fine dust is characterized by their micrometre size PM10 to PM2.5.
- Water pollution, by the release of waste products and contaminants into surface runoff into river drainage systems, leaching into groundwater, liquid spills, wastewater discharges, eutrophication and littering.

- Soil contamination occurs when chemicals are released by spill or underground leakage. Among the most significant soil contaminants are hydrocarbons, heavy metals, MTBE, herbicides, pesticides and chlorinated hydrocarbons.
- Littering
- Radioactive contamination, resulting from 20th century activities in atomic physics, such as nuclear power generation and nuclear weapons research, manufacture and deployment.
- Noise pollution, which encompasses roadway noise, aircraft noise, industrial noise as well as high-intensity sonar.
- Light pollution, includes light trespass, over-illumination and astronomical interference.
- Visual pollution, which can refer to the presence of overhead power lines, motorway billboards, scarred landforms (as from strip mining), open storage of trash or municipal solid waste.
- Thermal pollution, is a temperature change in natural water bodies caused by human influence, such as use of water as coolant in a power plant.

Pollutants

A pollutant is a waste material that pollutes air, water or soil. Three factors determine the severity of a pollutant: its chemical nature, the concentration and the persistence

Sources and Causes

Air pollution comes from both natural and manmade sources. Though globally manmade pollutants from combustion, construction, mining, agriculture and warfare are increasingly significant in the air pollution equation.

Motor vehicle emissions are one of the leading causes of air pollution. China, United States, Russia, Mexico, and Japan are the world leaders in air pollution emissions. Principal stationary pollution sources include chemical plants, coal-fired power plants, oil refineries, petrochemical plants, nuclear waste disposal activity, incinerators, large livestock farms (dairy cows, pigs, poultry, etc.), PVC factories, metals production factories, plastics factories, and other heavy industry. Agricultural air pollution comes from contemporary practices which include clear felling and burning of natural vegetation as well as spraying of pesticides and herbicides Some of the more common soil contaminants are chlorinated hydrocarbons (CFH), heavy metals (such as chromium, cadmium—found in rechargeable batteries, and lead—found in lead paint, aviation fuel and still in some countries, gasoline), MTBE, zinc, arsenic and benzene. In 2001 a series of press reports culminating in a book called Fateful Harvest unveiled a widespread practice of recycling industrial byproducts into fertilizer, resulting in the contamination of the soil with various metals. Ordinary municipal landfills are the source of many chemical substances entering the soil environment (and often groundwater), emanating from the wide variety of refuse accepted, especially substances illegally discarded there, or from pre-1970 landfills that may have been subject to little control in the U.S. or EU. There have also been some unusual releases of polychlorinated dibenzodioxins, commonly called dioxins for simplicity, such as TCDD. Pollution can also be the consequence of a natural disaster. For example, hurricanes often involve water contamination from sewage, and petrochemical spills from ruptured boats or automobiles. Larger scale and environmental damage is not uncommon when coastal oil rigs or refineries are involved. Some sources of pollution, such as nuclear power plants or oil tankers, can produce widespread and potentially hazardous releases when accidents occur. In the case of noise pollution the dominant source class is the motor vehicle, producing about ninety percent of all unwanted noise worldwide.

EFFECTS

Human Health

Adverse air quality can kill many organisms including humans. Ozone pollution can cause respiratory disease, cardiovascular disease, throat inflammation, chest pain, and congestion. Water pollution causes approximately 14,000 deaths per day, mostly due to contamination of drinking water by untreated sewage in developing countries. Oil spills can cause skin irritations and rashes. Noise pollution induces hearing loss, high blood pressure, stress, and sleep disturbance. Mercury has been linked to developmental deficits in children and neurologic symptoms. Lead and other heavy metals have been shown to cause neurological problems. Chemical and radioactive substances can cause cancer and as well as birth defects.

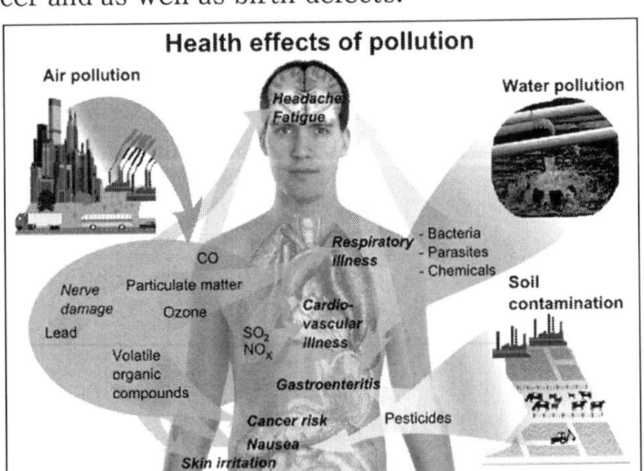

Ecosystems

- Sulphur dioxide and nitrogen oxides can cause acid rain which lowers the pH value of soil.
- Nitrogen oxides are removed from the air by rain and fertilise land which can change the species composition of ecosystems.
- Soil can become infertile and unsuitable for plants. This will affect other organisms in the food web.
- Smog and haze can reduce the amount of sunlight received by plants to carry out photosynthesis and leads to the production of tropospheric ozone which damages plants.
- Invasive species can out compete native species and reduce biodiversity. Invasive plants can contribute debris and biomolecules (allelopathy) that can alter soil and chemical compositions of an environment, often reducing native species competitiveness.
- Biomagnification describes situations where toxins (such as heavy metals) may pass through trophic levels, becoming exponentially more concentrated in the process.
- Carbon dioxide emissions cause ocean acidification, the ongoing decrease in the pH of the Earth's oceans as CO_2 becomes dissolved.
- The emission of greenhouse gases leads to global warming which affects ecosystems in many ways.

REGULATION AND MONITORING

To protect the environment from the adverse effects of pollution, many nations worldwide have enacted legislation to regulate various types of pollution as well as to mitigate the adverse effects of pollution.

Pollution Control

Pollution control is a term used in environmental management. It means the control of emissions and effluents into air, water or soil. Without pollution control, the waste products from consumption, heating, agriculture, mining, manufacturing, transportation and other human activities, whether they accumulate or disperse, will degrade the environment. In the hierarchy of controls, pollution prevention and waste minimization are more desirable than pollution control.

AIR POLLUTION

Air pollution is thc introduction of chemicals, particulate matter, or biological materials that cause harm or discomfort to humans or other living organisms, or damages the natural environment, into the atmosphere. The atmosphere is a complex, dynamic natural gaseous system that is essential to support life on planet Earth. Stratospheric ozone depletion due to air pollution has long been recognized as a threat to human health as well as to the Earth's ecosystems.

POLLUTANTS

An air pollutant is known as a substance in the air that can cause harm to humans and the environment. Pollutants can be in the form of solid particles, liquid droplets, or gases. In addition, they may be natural or man-made.

Pollutants can be classified as either primary or secondary. Usually, primary pollutants are substances directly emitted from a process, such as ash from a volcanic eruption, the carbon monoxide gas from a motor vehicle exhaust or sulfur dioxide released from factories.

Secondary pollutants are not emitted directly. Rather, they form in the air when primary pollutants react or interact. An important example of a secondary pollutant is ground level ozone - one of the many secondary pollutants that make up photochemical smog.

Note that some pollutants may be both primary and secondary: that is, they are both emitted directly and formed from other primary pollutants.

About 4 percent of deaths in the United States can be attributed to air pollution, according to the Environmental Science Engineering Programme at the Harvard School of Public Health.

Major Primary Pollutants Produced by Human Activity Include

- Sulfur oxides (SO_x): Especially sulfur dioxide, a chemical compound with the formula SO_2. SO_2 is produced by volcanoes and in various industrial processes. Since coal and petroleum often contain sulfur compounds, their combustion generates sulfur dioxide. Further oxidation of SO_2, usually in the presence of a catalyst such as NO_2, forms H_2SO_4, and thus acid rain. This is one of the causes for concern over the environmental impact of the use of these fuels as power sources.
- Nitrogen oxides (NO_x): Especially nitrogen dioxide are emitted from high temperature combustion. Can be seen as the brown haze dome above or plume downwind of cities.Nitrogen dioxide is the chemical compound with the formula NO_2. It is one of the several nitrogen oxides. This reddish-brown toxic gas has a characteristic sharp, biting odour. NO_2 is one of the most prominent air pollutants.
- Carbon monoxide - is a colourless, odourless, non-irritating but very poisonous gas. It is a product by incomplete combustion of fuel such as natural gas, coal or wood. Vehicular exhaust is a major source of carbon monoxide.
- Carbon dioxide (CO_2): A greenhouse gas emitted from combustion but is also a gas vital to living organisms. It is a natural gas in the atmosphere.

- Volatile organic compounds: VOCs are an important outdoor air pollutant. In this field they are often divided into the separate categories of methane (CH_4) and non-methane (NMVOCs). Methane is an extremely efficient greenhouse gas which contributes to enhanced global warming. Other hydrocarbon VOCs are also significant greenhouse gases via their role in creating ozone and in prolonging the life of methane in the atmosphere, although the effect varies depending on local air quality. Within the NMVOCs, the aromatic compounds benzene, toluene and xylene are suspected carcinogens and may lead to leukemia through prolonged exposure. 1,3-butadiene is another dangerous compound which is often associated with industrial uses.
- Particulate matter: Particulates, alternatively referred to as particulate matter (PM) or fine particles, are tiny particles of solid or liquid suspended in a gas. In contrast, aerosol refers to particles and the gas together. Sources of particulate matter can be man made or natural.
- Some particulates occur naturally, originating from volcanoes, dust storms, forest and grassland fires, living vegetation, and sea spray. Human activities, such as the burning of fossil fuels in vehicles, power plants and various industrial processes also generate significant amounts of aerosols.
- Averaged over the globe, anthropogenic aerosols—those made by human activities—currently account for about 10 percent of the total amount of aerosols in our atmosphere. Increased levels of fine particles in the air are linked to health hazards such as heart disease, altered lung function and lung cancer.
- Toxic metals, such as lead, cadmium and copper.
- Chlorofluorocarbons (CFCs): harmful to the ozone layer emitted from products currently banned from use.
- Ammonia (NH_3): Emitted from agricultural processes. Ammonia is a compound with the formula NH_3. It is normally encountered as a gas with a characteristic pungent odour. Ammonia contributes significantly to the nutritional needs of terrestrial organisms by serving as a precursor to foodstuffs and fertilizers. Ammonia, either directly or indirectly, is also a building block for the synthesis of many pharmaceuticals. Although in wide use, ammonia is both caustic and hazardous.
- Odours: Such as from garbage, sewage, and industrial processes
- Radioactive pollutants: Produced by nuclear explosions, war explosives, and natural processes such as the radioactive decay of radon.

Secondary Pollutants Include

- Particulate matter formed from gaseous primary pollutants and compounds in photochemical smog.Smog is a kind of air pollution; the word "smog" is a portmanteau of smoke and fog. Classic smog results from large amounts of coal burning in an area caused by a mixture of smoke and sulfur dioxide. Modern smog does not usually come from coal but from vehicular and industrial emissions that are acted on in the atmosphere by sunlight to form secondary pollutants that also combine with the primary emissions to form photochemical smog.
- Ground level ozone (O_3) formed from NO_x and VOCs. Ozone (O_3) is a key constituent of the troposphere (it is also an important constituent of certain regions of the stratosphere commonly known as the Ozone layer). Photochemical and chemical reactions involving it drive many of the chemical processes that occur in the atmosphere by day and by night. At abnormally high concentrations brought about by human activities (largely the combustion of fossil fuel), it is a pollutant, and a constituent of smog.
- Peroxyacetyl nitrate (PAN) - similarly formed from NO_x and VOCs.

Minor Air Pollutants Include

- A large number of minor hazardous air pollutants. Some of these are regulated in USA under the Clean Air Act and in Europe under the Air Framework Directive.
- A variety of persistent organic pollutants, which can attach to particulate matter.

Persistent organic pollutants (POPs) are organic compounds that are resistant to environmental degradation through chemical, biological, and photolytic processes. Because of this, they have been observed to persist in the environment, to be capable of long-range transport, bioaccumulate in human and animal tissue, biomagnify in food chains, and to have potential significant impacts on human health and the environment.

Sources

Sources of air pollution refer to the various locations, activities or factors which are responsible for the releasing of pollutants in the atmosphere. These sources can be classified into two major categories which are:

Anthropogenic sources (human activity) mostly related to burning different kinds of fuel.

- "Stationary Sources" include smoke stacks of power plants, manufacturing facilities (factories) and waste incinerators, as well as furnaces and other types of fuel-burning heating devices.

- "Mobile Sources" include motor vehicles, marine vessels, aircraft and the effect of sound etc.
- Chemicals, dust and controlled burn practices in agriculture and forestry management. Controlled or prescribed burning is a technique sometimes used in forest management, farming, prairie restoration or greenhouse gas abatement. Fire is a natural part of both forest and grassland ecology and controlled fire can be a tool for foresters. Controlled burning stimulates the germination of some desirable forest trees, thus renewing the forest.
- Fumes from paint, hair spray, varnish, aerosol sprays and other solvents.
- Waste deposition in landfills, which generate methane. Methane is not toxic; however, it is highly flammable and may form explosive mixtures with air. Methane is also an asphyxiant and may displace oxygen in an enclosed space. Asphyxia or suffocation may result if the oxygen concentration is reduced to below 19.5per cent by displacement
- Military, such as nuclear weapons, toxic gases, germ warfare and rocketry.

Natural Sources

- Dust from natural sources, usually large areas of land with little or no vegetation.
- Methane, emitted by the digestion of food by animals, for example cattle.
- Radon gas from radioactive decay within the Earth's crust. Radon is a colourless, odourless, naturally occurring, radioactive noble gas that is formed from the decay of radium. It is considered to be a health hazard.Radon gas from natural sources can accumulate in buildings, especially in confined areas such as the basement and it is the second most frequent cause of lung cancer, after cigarette smoking.
- Smoke and carbon monoxide from wildfires.
- Volcanic activity, which produce sulfur, chlorine, and ash particulates.

Emission Factors

Air pollutant emission factors are representative values that attempt to relate the quantity of a pollutant released to the ambient air with an activity associated with the release of that pollutant.

These factors are usually expressed as the weight of pollutant divided by a unit weight, volume, distance, or duration of the activity emitting the pollutant (*e.g.*, kilograms of particulate emitted per megagram of coal burned). Such factors facilitate estimation of emissions from various sources of air pollution. In most cases, these factors are simply averages of all available data of acceptable quality, and are generally assumed to be representative of long-term averages.

The United States Environmental Protection Agency has published a compilation of air pollutant emission factors for a multitude of industrial sources. The United Kingdom, Australia, Canada and other countries have published similar compilations, as has the European Environment Agency.

AIR POLLUTION EFFECTS

We can distinguish between short-term acute effects and long-term chronic effects of air pollution. Acute effects may be brought on, for instance, by pollution episodes, while chronic effects may generally affect a greater proportion of the population. It is important to remember that air pollutants are transboundary, *i.e.* they know no borders and travel easily from their sources towards other locations spreading pollution throughout the world. The respiratory system is by far the most important route for entry of air pollutants into the body and thus becomes their primary target. Air pollution affects virtually everyone but is considered to affect poorer people more than wealthier ones.

AIR POLLUTION EFFECTS ON HUMANS

Air Pollution Effects: Sulfur Dioxide

Sulfur dioxide is a colourless gas with a pungent, suffocating odour. SO_2 is corrosive to organic materials and it irritates the eyes, nose and lungs; therefore it is quite a dangerous air pollutant. Sulfur is contained within all fossil fuels, and is released in the form of sulfur dioxide during fossil fuel combustion.

Because of the widespread use of fossil fuels, sulfur dioxide is among the most common air pollutants produced in every part of the planet. Sulfur dioxide may often act in synergy with other pollutants to produce certain air pollution effects. Let's summarize some sulfur dioxide pollution effects on human health:

- Irritation of eyes, nose, throat; damage to lungs when inhaled
- Acute and chronic asthma
- Bronchitis and emphysema
- Lung cancer

Air Pollution Effects: Nitrogen Dioxide

Nitrogen dioxide is a gas of reddish-brown colour with a distinct sharp, biting odour. It is often analysed in conjunction with another nitrogen gas–nitric oxide. Together these two gases are referred to as NO_x. Combustion of fossil fuels always produces both NO_2 and NO. But almost 90per cent of the NO_x combustion product is in the form of NO which is then oxidized to NO_2 in the air. NO_x can destroy organic matter, ex. human tissue. Exposure to high concentrations of NO_x can make living organisms more susceptible to bacterial infections and lung cancer. Just like other pollutants, nitrogen dioxide affects people with existing medical conditions more severely than healthy people. For example, asthma sufferers may experience enhanced sensitivity after short-term NO_2 exposure as compared to those without any asthmatic problems. Another group at higher risk is children. For instance, children aged 12 and younger who are exposed to NOx suffer more respiratory illness than the ones who are not exposed.

Depending on different NO_2 concentrations in the air, nitrogen dioxide pollution effects may be as follows:

- Increased incidence of respiratory illness
- Increased airway resistance
- Damage to lung tissue
- Chronic obstructive pulmonary disease, or COPD
- Emphysema
- Pulmonary edema
- Infant and cardiovascular death

It is important to note that nitrogen dioxide is a major component of the photochemical smog and thus a contributor to the formation of ozone which is another serious air pollutant

Air Pollution Effects: Carbon Monoxide

Carbon monoxide is an extremely toxic gas which affects the ability of the body to receive oxygen. Hemoglobin which transports oxygen in the blood is bound by carbon monoxide, which leads to the shortage of oxygen in the body. Carbon monoxide is the most common type of fatal poisoning in many countries around the world. Unlike many other air pollutants, carbon monoxide does not directly affect eyes, nasal passages or lungs.

Exposure to carbon monoxide may lead to:

- Toxicity of the central nervous system and heart
- Headaches, dizziness, nausea and unconsciousness
- Loss of vision
- Decreased muscular coordination
- Abdominal pain
- Severe effects on the baby of a pregnant woman
- Impaired performance on simple psychological tests and arithmetic; loss of judgement of time
- In cases of prolonged exposure to high CO concentrations, unconsciousness, convulsions and death would occur

Carbon monoxide pollution effects are exacerbated in weaker people, specifically in those with heart and lung diseases. What makes this gas even more dangerous is the fact that it is invisible and odourless, and thus not immediately detectable as a hazard.

Air Pollution Effects: Ozone Effects

Ozone is a poisonous gas with a sharp and cold odour. It can be found in the stratosphere where it occurs naturally and plays a beneficial role by protecting the Earth from ultraviolet sunlight; and in the troposphere where it occurs naturally and also forms part of the human-caused photochemical smog. It is of course the tropospheric ozone that we are interested in as an important air pollutant. The chemical reaction between nitrogen oxides and volatile organic compounds in the presence of sunlight results in the photochemical smog; the tropospheric ozone is an end product of this reaction and a component of the smog itself. Because the photochemical smog requires a lot of sunshine to form, it occurs mostly in sunny and heavily polluted places, such as Los Angeles, Mexico City and even Athens. Ozone's main "victim" within the human body is its respiratory system. Once in the lungs, ozone burns through cell walls. The immune system fails to protect the lungs because ozone pushes the defensive cells back. As cellular fluid starts seeping into the lungs, breathing becomes rapid, shallow and painful. Exposure to ozone over long periods of time leads to a stiffening of the lungs and a reduced ability to breathe. As an example, a study conducted in California in the 1980s shows that children living in ozone-polluted areas have smaller than normal lungs and adults lose up to 75per cent of their lung capacity.

So, exposure to the tropospheric ozone may cause:

- Burning nose and watering eyes
- Tightening of the chest
- Coughing, wheezing and throat irritation
- Rapid, shallow, painful breathing
- Susceptibility to respiratory infections
- Inflammation and damage to the lining of the lungs
- Aggravation of asthma
- Fatigue
- Cancer

Air Pollution Effects: Ammonia Effects

Ammonia is a colourless, pungent, hazardous caustic gas composed of nitrogen and hydrogen. Though ammonia is used for different applications in many sectors, agriculture is its largest consumer and producer. Livestock farming, animal waste and fertilizer application are the biggest sources of atmospheric ammonia emissions within the agricultural sector. Gaseous ammonia is a dangerous air pollutant. Breathing in large amounts can cause death.

Ammonia Pollution Effects

On the respiratory system:

- Nose and throat irritation and burns
- Swelling of the throat and airways; airways destruction

- Pulmonary edema
- Chronic lung disease
- Cough
- Asthma
- Lung fibrosis
- Inhaling large amounts of ammonia can be fatal

On the skin and eyes:

- Skin burns
- Skin conditions, ex. dermatitis
- Burning sensation in the eyes
- Ulceration and perforation of the cornea; blindness
- Cataracts and glaucoma

Air Pollution Effects: Volatile Organic Compounds

Volatile organic compounds are defined as organic compounds which easily evapourate and enter the atmosphere. VOCs are important pollutants for two reasons. First, they are precursors to the formation of ozone; second, they include compounds which are carcinogenic and mutagenic in their own right. Typical VOCs include propane, benzene, ethanol, methanol, ether, carbon tetrachloride and vinyl chloride; substances such as petrol and resins contain many individual VOCs, and many others are produced during combustion processes. Toxicity of some VOCs and ensuing health effects are no doubt issues of serious concern. For example, exposure to benzene and 1,3-butadiene is a suspected cause of around 10per cent of leukemia incidence in the UK. Exposure to toluene – another dangerous VOC – may lead to the dysfunction of the central nervous system resulting in behavioural problems, memory loss and disturbance of the circadian rhythm; toluene is also suspected to cause cancer.

Some other VOCs, ex. carbon tetrachloride and PCBs, are believed to produce abnormal changes in fetus development and consequently lead to birth defects. Carbon tetrachloride also leads to liver damage. Vinyl chloride causes Raynaud's phenomenon, necrosis of the small bones of the hand, liver damage, and a rare, highly malignant tumor of the liver.

Benzene pollution effects on human health:

- Short-term breathing of low levels of benzene can cause drowsiness, dizziness, rapid heart rate, headaches, tremors, confusion and unconsciousness; exposure to high levels of benzene may result in death
- *Chronic effects*:
  - — Damage to the bone marrow
  - — Decrease in red blood cells
  - — Excessive bleeding and depression of the immune system increasing the risk of infection
  - — Leukemia

Butadiene pollution effects on human health:

- Short-term exposure to high levels of 1,3-butadiene may lead to distorted blurred vision, vertigo, general tiredness, decreased blood pressure, headache, nausea, decreased pulse rate, and fainting
- Long-term exposure to high levels can increase the risk of cardiovascular diseases and cancer

Let us summarize some effects of VOCs on human health:

- Tiredness, vertigo, drowsiness, dizziness, nausea, confusion, unconsciousness
- Anemia
- Bone marrow damage
- Liver damage
- *Dysfunction of the central nervous system*:
  - — Behavioural problems
  - — Memory loss
  - — Disturbance of the circadian rhythm
- Cardiovascular diseases
- Cancer; specifically leukemia
- Abnormal changes in fetus development, birth defects
- VOCs also contribute to sick building syndrome indoors
- As facilitators in ozone formation, VOCs may indirectly contribute to respiratory problems

AIR POLLUTION EFFECTS: AIRBORNE PARTICLES

Airborne particles are tiny fragments of solid or liquid nature suspended in the air. Particles may be primary – when emitted into the atmosphere directly by sources, or secondary – when formed in the atmosphere through the interaction of primary pollutants. For example, sulfur oxides and nitrogen oxides are primary pollutants which may transform in the air into sulfates and nitrates.

Dust particles are solid particles between 1 and 100 ìm in diametre; fumes, or smoke, are solid particles less than 1 ìm in diametre. Experts believe that dust is the most damaging among all widely measured air pollutants. Smaller dust particles are more dangerous than larger ones because they can penetrate deep into the lungs being deposited on areas where the body's natural clearance mechanisms such as coughing cannot remove them. Combustion of fossil fuels by road transport, power plants etc. is a major source of particulate air pollution. Particles may come in a whole variety of chemical compositions including heavy metals such as cadmium, mercury and lead. They often act in synergy with other air pollutants, ex. sulfur dioxide, to produce negative health effects in humans.

As particulate pollution levels rise, it is common to see an increase in hospital admissions for:

- Stuffy noses, sinusitis
- Sore throats
- Wet cough, dry cough, phlegm
- Head colds
- Burning eyes
- Wheezing; shortness of breath
- Lung disease
- Chest discomfort or pain

As with other pollutants, children are of course more susceptible to particulate pollution.

Specific children's disorders caused by airborne particles may include:

- Infant death
- Low birth weight
- Reduced lung function

Here is a summary of the particulate pollution effects provided by the US Environmental Protection Agency:

- Increased respiratory symptoms
- Decreased lung function
- Aggravated asthma
- Chronic bronchitis
- Irregular heartbeat
- Nonfatal heart attacks
- Premature death in people with heart or lung disease

Indeed, particulate pollution affects both developed and developing countries:

— For example, a major study conducted in the US between 1982 and 1989 found that people living in American cities with the highest levels of particulate air pollution were 15per cent to 17per cent more likely to die one to three years early

— In another example, particulate air pollution is linked to nearly 350,000 premature deaths in China and India every year

AIR POLLUTION EFFECTS ON ANIMALS

We don't know a lot about air pollution effects on animals.

- Probably one of the best examples here is that of acid rain and how it affects freshwater animal life.

Sulfur dioxide and nitrogen dioxide are transformed in the atmosphere to produce acid compounds–sulfuric and nitric acids. These compounds then fall back on to the ground as particulates or raindrops—in other words, acid rain. So acid rain also falls on streams and lakes, acidifies them and destroys fish life in these freshwater ecosystems. For example, in Sweden acid rain made over 18,000 lakes so acidic that all the fish died out. Salmon species appear to be particularly sensitive to acidity. Some other populations of animals in Europe and North America that have also been declining due to acid rain are brown trout, mayfly larvae, beetle larvae, mollusks, and aquatic bird species.

- Pollution may also affect animals through plants on which they feed.

For example, pea aphids feed on pea plants exposed to sulfur dioxide in the air. High exposure to sulfur dioxide negatively affects the health of the pea plants, and therefore, the health of the aphids as well.

Some other examples of air pollution effects on animals:

- Excessive ultraviolet radiation coming from the sun through the ozone layer in the upper atmosphere which is eroded by some air pollutants, may cause skin cancer in wildlife
- Tropospheric ozone may damage lung tissues of animals

It is also probably logical to assume that many higher order animals experience air pollution effects similar to those experienced by humans.

AIR POLLUTION EFFECTS ON FORESTS, TREES AND PLANTS

Air pollution can have both long-term and short-term effects on plants.

- Physical injury to leaves is the immediate effect of air pollution on plants. Here is how leaves are affected by different air pollutants:
  - — Ozone produces a speckle of brown spots, which appear on the flat areas of leaf between the veins
  - — *Sulfur dioxide*: Larger bleached-looking areas
  - — *Nitrogen dioxide*: *I*rregular brown or white collapsed lesions on intercostal tissue and near the leaf edge
  - — *Ammonia*: Unnatural green appearance with tissue drying out
- Of all main air pollutants, sulfur dioxide often comes up as the one that most negatively affects plants and trees.

Here is a very illustrative example of how destructive sulfur dioxide can be to vegetation. The development of ore deposits in Canada in the middle of the 20 th century was a major source of sulfur dioxide emissions in high concentrations. Sulfur emissions of one specific iron smelting plant in Wawa, Ontario, caused the destruction of large tracts of native boreal forest in a belt stretching some 30 km from the smelter.

Another good example is that of the Black Triangle – the area where the borders of Germany, Poland and the Czech Republic meet, with significant brown coal deposits. Industrial activity in this area led to high sulfur emissions which resulted in devastation of many square kilometres of forest, particularly above an altitude of 750 m. However, it has been noted that sulfur dioxide concentrations in the atmosphere have recently been significantly reduced almost throughout the world and do not carry such dramatic effects as they used to during the 20th century.

Lichens are considered to be most sensitive to sulfur dioxide. During the period of high levels of sulfur pollution, large parts of Europe lost many species of lichen and became known as "lichen deserts". But as sulfur pollution levels dropped, many lichen species re-appeared. Sulfur dioxide may also affect higher plants, including wild species, crops and trees.

These effects may be:

- Cell metabolism disruption
- Leaf injury and loss
- Reduced growth and reproduction
- Increase in susceptibility of plants to attacks by insect herbivores
- Nitrogen dioxide, another air pollutant, may act in synergy with sulfur dioxide to produce a negative effect on plants' photosynthesis.
- Tropospheric ozone can prevent plant respiration by blocking stomata and negatively affecting plants' photosynthesis rates which will stunt plant growth; ozone can also decay plant cells directly by entering stomata.
- Particles, just like ozone, often affect plants and trees via blocking of leaf stomata through which plants undertake the gas exchange necessary for photosynthesis and respiration.

Andrew Farmer notes that:

- In sufficient quantities, dusts may form a smothering layer on leaves, reducing light and hence lowering photosynthetic rates. Many dusts are inert and so only act by shading.

However, some dusts are also chemically active. Thus cement dust will also dissolve leaf tissue, resulting in additional injury. Coal dust may also contain toxic compounds.

Dusts may also affect ecosystems through their action on soil. Thus the alkaline chemistry of limestone dusts can raise the soil pH of acid and neutral habitats, resulting in the loss of plant and animal species."

So, particulate air pollution effects on plants and trees may be as follows:

- Blocked stomata
- Increased leaf temperature
- Reduced photosynthesis
- Reduced fruit set, leaf growth, pollen growth
- Reduced tree growth
- Leaf necrosis and chlorosis, bark peeling
- Acid rain severely affects trees and plants as well.

It can kill trees, destroy the leaves of plants, can infiltrate soil by making it unsuitable for purposes of nutrition and habitation. It is also associated with the reduction in forest and agricultural yields.

AIR POLLUTION EFFECTS ON WIDER ENVIRONMENT

The effects of acid rain have been known for a long time. Though experts admit that the problem of acid rain is generally under control, it's still worth reiterating the effects brought on by acid rain on the wider environment.

Acid rain has adverse effects on:

- Forests and other vegetation
- Freshwater lakes and streams destroying aquatic life
- Soil
- Buildings and materials

Air pollution also has a negative impact on visibility.

ECONOMIC LOSSES AS AIR POLLUTION EFFECTS

Apart from direct health-related and other environmental issues, air pollution brings with it economic losses as well.

Some of the economic losses caused by air pollution are as follows:

- Direct medical losses
- Lost income from being absent from work
- Decreased productivity
- Travel time losses due to reduced visibility
- Losses from repair of damage to buildings
- Increased costs of cleaning
- Losses due to damage to crops and plants

For example, experts calculated that California is losing $28 billion every year in health care costs, school absences, missed work and lost income potential from premature deaths, because of air pollution in Southern California and in the San Joaquin Valley.

LAND POLLUTION

Land Pollution is the degradation of Earth's land surfaces often caused by human activities and their misuse of land resources. It occurs when waste is not disposed properly.Haphazard disposal of urban and industrial wastes, exploitation of minerals, and improper use of soil by inadequate agricultural practices are a few factors. Urbanization and industrialisation are major causes of land pollution.

The Industrial Revolution set a series of events into motion which destroyed natural habitats and polluted the environment, causing diseases in both humans and animals.

CAUSES

The increase in the concentration of population in cities, along with the internal combustion engine, led to the increased number of roads and all the infra structure that goes with them. Roads cause visual, noise, light, air and water pollution, in addition to land pollution. The visual and noise areas are obvious, however light pollution is becoming more widely recognised as a problem. From outer space, large cities can be picked out at night by the glow of their lighting, so city dwellers seldom experience total darkness.

The contribution of vehicular traffic to air pollution is dealt with in another article, but, suffice to say that sulfur dioxide, nitrogen oxide and carbon monoxide are the main culprits. Water pollution is caused by the run off from roads of oil, salt and rubber residue, which enter the water courses and may make conditions unsuitable for certain organisms to live.

As the demand for food has grown very high, there is an increase in field size and mechanization. The increase in field size makes it economically viable for the farmer but results in loss of habitat and shelter for wildlife, as hedgerows and copses disappear. When crops are harvested, the naked soil is left open to wind after the heavy machinery has compacted it.

Another consequence of more intensive agriculture is the move to monoculture. This is unnatural, it depletes the soil of nutrients, allows diseases and pests to spread and, in short, brings into play the use of chemical substances foreign to the environment.

Pesticides

A pesticide is a substance or mixture of substances used to kill a pest. A pesticide may be a chemical substance, biological agent (such as a virus or bacteria), antimicrobial, disinfectant or device used against any pest. Pests include insects, plant pathogens, weeds, molluscs, birds, mammals, fish, nematodes (roundworms) and microbes that compete with humans for food, destroy property, spread or are a vector for disease or cause a nuisance.

Although there are benefits to the use of pesticides, there are also drawbacks, such as potential toxicity to humans and other animals. Pesticides are used to control organisms which are considered harmful. For example, they are used to kill mosquitoes that can transmit potentially deadly diseases like west nile virus, yellow fever, and malaria. They can also kill bees, wasps or ants that can cause allergic reactions. Insecticides can protect animals from illnesses that can be caused by parasites such as fleas. Pesticides can prevent sickness in humans that could be caused by mouldy food or diseased produce.

Herbicides

Herbicides are used to kill weeds, especially on pavements and railways. They are similar to auxins and most are biodegradable by soil bacteria. However one group derived from trinitrophenol (2:4 D and 2:4:5 T) have the impurity dioxin, which is very toxic and causes fatality even in low concentrations.

It also causes spontaneous abortions, haemorrhaging and cancer. Agent Orange (50per cent 2:4:5 T) was used as a defoliant in Vietnam. Eleven million gallons were used and children born since then to American soldiers who served in this conflict, have shown increased physical and mental disabilities compared to the rest of the population. It affects the head of the sperm and the chromosomes inside it.

Another herbicide, much loved by murder story writers, is Paraquat. It is highly toxic but it rapidly degrades in soil due to the action of bacteria and does not kill soil fauna.

Insecticides

Insecticides are used to rid farms of pests which damage crops. The insects damage not only standing crops but also stored ones and in the tropics it is reckoned that one third of the total production is lost during food storage.

As with fungicides, the first insecticides used in the nineteenth century were inorganic *e.g.* Paris Green and other compounds of arsenic. Nicotine has also been used since the late eighteenth century. There are now two main groups of synthetic insecticides.

Organochlorines

Organochlorines include DDT, Aldrin, Dieldrin and BHC. They are cheap to produce, potent and persistent. DDT was used on a massive scale from the 1930s, with a peak of 72,000 tonnes used 1970. Then usage fell as the harmful environmental effects were realized. It was found worldwide in fish and birds and was even discovered in the snow in the Antarctic. It is only slightly soluble in water but is very soluble in the bloodstream.

It affects the nervous and endocrine systems and causes the eggshells of birds to lack calcium causing them to be easily breakable. It is thought to be responsible for the decline of the numbers of birds of prey like ospreys and peregrine falcons in the 1950s - they are now recovering.

As well as increased concentration via the food chain, it is known to enter via permeable membranes, so fish get it through their gills. As it has low water solubility, it tends to stay at the water surface, so organisms that live there are most affected. DDT found in fish that formed part of the human food chain caused concern, but the levels found in the liver, kidney and brain tissues was less than 1ppm and in fat was 10 ppm which was below the level likely to cause harm. However, DDT was banned in Britain and America to stop the further build up of it in the food chain. The USA exploited this ban and sold DDT to developing countries, who could not afford the expensive replacement chemicals and who did not have such stringent regulations governing the use of pesticides.

Some insects have developed a resistance to insecticides - *e.g.* the which carries malaria.

Organophosphates

Organophosphates, *e.g.* parathion, methyl parathion and about 40 other insecticides are available nationally. Parathion is highly toxic, methyl-parathion is less so and Malathion is generally considered safe as it has low toxicity and is rapidly broken down in the mammalian liver. This group works by preventing normal nerve transmission as cholinesterase is prevented from breaking down the transmitter substance acetylcholine, resulting in uncontrolled muscle movements. Entry of a variety of pesticides into our water supplies causes concern to environmental groups, as in many cases the long term effects of these specific chemicals is not known.

Restrictions came into force in July 1985 and were so frequently broken that in 1987, formal proceedings were taken against the British government. Britain is still the only European state to use Aldrin and organochlorines, although it was supposed to stop in 1993. East Anglia has the worst record for pesticide contamination of drinking water. Of the 350 pesticides used in Britain, only 50 can be analyzed, which is worrying for the global community.

MINING

Modern mining projects leave behind distrupted communities, damages landscapes, and polluted water. Mining also affects ground and surface waters, the aquatic life, vegetation, soils, animals, and the human health. Acid mine drainage can cause damage to streams which in return can kill aquatic life. The vast variety of toxic chemicals released by mining activities can harm animals and aquatic life as well as their habitat. The average mine disturbs over a thousand acres of land.

BURIAL

Burial is the technique used by Jews, Muslims, Christians and other religions with Abrahamic influence, to dispose off the corpse of dead humans and animals. This process leads to regular soil erosion due to loosening of soil.

Also, the decomposing fluids act as poisonous herbicides, pesticides and may even lead to epidemics in surrounding areas. It leads to soil pollution, soil erosion and even water pollution.

CONSTRUCTION

Construction often puts sediments in rivers and bodies of water. By doing this, natural water filters are damaged. Natural water filters help break down many pollutants before they reach other water bodies. Some harmful chemicals that may run off with water and sediments from construction sites are oils, debris, and paint.

This can cause damage to soil, aquatic life, and promote hazordous chemicals to get into drinking water. California Integrated Waste Management Board provides more "green building basics" that educates readers about healthy construction.

Increased Waste Disposal

In Scotland in 1993, 14 million tons of waste was produced. 100,000 tons was special waste and 260,000 tons was controlled waste from other parts of

Britain and abroad. 45per cent of the special waste was in liquid form and 18per cent was asbestos - radioactive waste was not included.

Of the controlled waste, 48per cent came from the demolition of buildings, 22 per cent from industry, 17 per cent from households and 13 per cent from business - only 3 per cent were recycled. 90 per cent of controlled waste was buried in landfill sites and produced 2 million tons of methane gas. 1.5 per cent was burned in incinerators and 1.5 per cent were exported to be disposed of or recycled. There are 748 disposal sites in Scotland.

Landfills produce leachate, which has to be recycled to keep favourable conditions for microbial activity, methane gas and some carbon dioxide.

There are very few vacant or derelict land sites in the north east of Scotland, as there are few traditional heavy industries or coal/mineral extraction sites. However some areas are contaminated by aromatic hydrocarbons (500 cubic meters).

The Urban Waste Water Treatment Directive allows sewage sludge to be sprayed onto land and the volume is expected to double to 185,000 tons of dry solids in 2005.

This has good agricultural properties due to the high nitrogen and phosphate content. In 1990/1991, 13 per cent wet weight was sprayed onto 0.13 per cent of the land, however this is expected to rise 15 fold by 2005. There is a need to control this so that pathogenic microorganisms do not get into water courses and to ensure that there is no accumulation of heavy metals in the top soil.

Impacts

A large percentage of material that is dumped into waters not only destroys habitats for animals but for humans as well. Below is a list of some impacts of land pollution..

- Smothered habitats, where aquatic animals feed and live, thus driving them out of a home.
- Reduced oxygen levels from decomposition of organic material dumped from drains.
- Reduced light penetration in the water column, from sediments and excessive algal growth, leading to oxygen depleted water. *Massively changed flow regimes, large fast flows can scour existing habitat and wash species downstream.
- Increased human health risks, from syringes, broken glass and cans washing onto creek banks and beaches.
- Reduced aesthetic appeal of areas from litter strewn on banks and beaches, thus reducing recreation and tourist appeal.
- Acid rain damages trees and other plants.

Reduce Land Pollution

There are many ways to help reduce land pollution. Some ideas are listed below:

- Use reusable materials.
- Do not litter.
- Recycle.
- Purchase products with little packaging to throw away (buy in bulk if possible).
- Use safer alternative pesticides.
- Buy organic grown fruits.
- Buy biogradeable products.

MARINE POLLUTION

Marine pollution occurs when harmful effects, or potentially harmful effects, can result from the entry into the ocean of chemicals, particles, industrial, agricultural and residential waste, or the spread of invasive organisms.

Most sources of marine pollution are land based. The pollution often comes from non-point sources such as agricultural runoff and wind blown debris.

Many potentially toxic chemicals adhere to tiny particles which are then taken up by plankton and benthos animals, most of which are either deposit or filter feeders. In this way, the toxins are concentrated upward within ocean food chains. Many particles combine chemically in a manner highly depletive of oxygen, causing estuaries to become anoxic.

When pesticides are incorporated into the marine ecosystem, they quickly become absorbed into marine food webs. Once in the food webs, these pesticides can cause mutations, as well as diseases, which can be harmful to humans as well as the entire food web.

Toxic metals can also be introduced into marine food webs. These can cause a change to tissue matter, biochemistry, behaviour, reproduction, and suppress growth in marine life. Also, many animal feeds have a high fish meal or fish hydrolysate content. In this way, marine toxins can be transferred to land animals, and appear later in meat and dairy products.

Pathways of Pollution

Bjorn Jenssen (2003) notes in his article, "Anthropogenic pollution may reduce biodiversity and productivity of marine ecosystems, resulting in reduction and depletion of human marine food resources".

There are many different ways to categorize, and examine the inputs of pollution into our marine ecosystems. Patin (n.d.) notes that generally there are three main types of inputs of pollution into the ocean: direct discharge of waste into the oceans, runoff into the waters due to rain, and pollutants that are released from the atmosphere. One common path of entry by contaminants to the sea are rivers. The Hudson in New York State and the Raritan in New Jersey, which empty at the northern and southern ends of Staten Island, are a source of mercury contamination of zooplankton (copepods) in the open ocean. The highest concentration in the filter-feeding copepods is not at the mouths of these rivers but 70 miles south, nearer Atlantic City, because water flows close to the coast. It takes a few days before toxins are taken up by the plankton.

Pollution is often classed as point source or non-point source pollution. Point source pollution occurs when there is a single, identifiable, and localized source of the pollution. An example is directly discharging sewage and industrial waste into the ocean.

Pollution such as this occurs particularly in developing nations. non-point source pollution occurs when the pollution comes from ill-defined and diffuse sources. These can be difficult to regulate. Agricultural runoff and wind blown debris are prime examples.

Pollution from Ships

Ships can pollute waterways and oceans in many ways. Oil spills can have devastating effects. While being toxic to marine life, polycyclic aromatic hydrocarbons (PAHs), the components in crude oil, are very difficult to clean up, and last for years in the sediment and marine environment.

Discharge of cargo residues from bulk carriers can pollute ports, waterways and oceans. In many instances vessels intentionally discharge illegal wastes despite foreign and domestic regulation prohibiting such actions. Ships create noise pollution that disturbs natural wildlife, and water from ballast tanks can spread harmful algae and other invasive species.

Meinesz believes that one of the worst cases of a single invasive species causing harm to an ecosystem can be attributed to a seemingly harmless jellyfish. Mnemiopsis leidyi, a species of comb jellyfish that spread so it now inhabits estuaries in many parts of the world. It was first introduced in 1982, and thought to have been transported to the Black Sea in a ship's ballast water.

The population of the jellyfish shot up exponentially and, by 1988, it was wreaking havoc upon the local fishing industry. "The anchovy catch fell from 204,000 tonnes in 1984 to 200 tons in 1993; sprat from 24,600 tonnes in 1984 to 12,000 tonnes in 1993; horse mackerel from 4,000 tonnes in 1984 to zero in 1993." Now that the jellyfish have exhausted the zooplankton, including fish larvae, their numbers have fallen dramatically, yet they continue to maintain a stranglehold on the ecosystem.

Invasive species can take over once occupied areas, facilitate the spread of new diseases, introduce new genetic material, alter underwater seascapes and jeopardize the ability of native species to obtain food. Invasive species are responsible for about $138 billion annually in lost revenue and management costs in the US alone.

PLASTIC DEBRIS

Marine debris is mainly discarded human rubbish which floats on, or is suspended in the ocean. Eighty percent of marine debris is plastic - a component that has been rapidly accumulating since the end of World War II. The mass of plastic in the oceans may be as high as one hundred million metric tons.

Discarded plastic bags, six pack rings and other forms of plastic waste which finish up in the ocean present dangers to wildlife and fisheries. Aquatic life can be threatened through entanglement, suffocation, and ingestion.

Fishing nets, usually made of plastic, can be left or lost in the ocean by fishermen. Known as ghost nets, these entangle fish, dolphins, sea turtles, sharks, dugongs, crocodiles, seabirds, crabs, and other creatures, restricting movement, causing starvation, laceration and infection, and, in those that need to return to the surface to breathe, suffocation.

Many animals that live on or in the sea consume flotsam by mistake, as it often looks similar to their natural prey. Plastic debris, when bulky or tangled, is difficult to pass, and may become permanently lodged in the digestive tracts of these animals, blocking the passage of food and causing death through starvation or infection.

Plastics accumulate because they don't biodegrade in the way many other substances do. They will photodegrade on exposure to the sun, but they do so properly only under dry conditions, and water inhibits this process. In marine environments, photodegraded plastic disintegrates into ever smaller pieces while remaining polymers, even down to the molecular level.

When floating plastic particles photodegrade down to zooplankton sizes, jellyfish attempt to consume them, and in this way the plastic enters the ocean food chain. Many of these long-lasting pieces end up in the stomachs of marine birds and animals, including sea turtles, and black-footed albatross.

Plastic debris tends to accumulate at the centre of ocean gyres. In particular, the Great Pacific Garbage Patch has a very high level of plastic particulate suspended in the upper water column. In samples taken in 1999, the mass of plastic exceeded that of zooplankton (the dominant animal life in the area) by a factor of six.

Toxic additives used in the manufacture of plastic materials can leach out into their surroundings when exposed to water. Waterborne hydrophobic pollutants collect and magnify on the surface of plastic debris, thus making plastic far more deadly in the ocean than it would be on land. Hydrophobic contaminants are also known to bioaccumulate in fatty tissues, biomagnifying up the food chain and putting pressure on apex predators.

Some plastic additives are known to disrupt the endocrine system when consumed, others can suppress the immune system or decrease reproductive rates. Floating debris can also absorb persistent organic pollutants from seawater, including PCBs, DDT and PAHs. Aside from toxic effects, when ingested some of these are mistaken by the animal brain for estradiol, causing hormone disruption in the affected wildlife.

TOXINS

Apart from plastics, there are particular problems with other toxins that do not disintegrate rapidly in the marine environment. Examples of persistent toxins are PCBs, DDT, pesticides, furans, dioxins and phenols. Heavy metals are metallic chemical elements that have a relatively high density and are toxic or poisonous at low concentrations. Examples are mercury, lead, nickel, arsenic and cadmium. Such toxins can accumulate in the tissues of many species of aquatic life in a process called bioaccumulation. They are also known to accumulate in benthic environments, such as estuaries and bay muds: a geological record of human activities of the last century.

Eutrophication

Eutrophication is an increase in chemical nutrients, typically compounds containing nitrogen or phosphorus, in an ecosystem. It can result in an increase in the ecosystem's primary productivity (excessive plant growth and decay), and further effects including lack of oxygen and severe reductions in water quality, fish, and other animal populations.

The biggest culprit are rivers that empty into the ocean, and with it the many chemicals used as fertilizers in agriculture as well as waste from livestock and humans. An excess of oxygen depleting chemicals in the water can lead to hypoxia and the creation of a dead zone.

Estuaries tend to be naturally eutrophic because land-derived nutrients are concentrated where runoff enters the marine environment in a confined channel. The World Resources Institute has identified 375 hypoxic coastal zones around the world, concentrated in coastal areas in Western Europe, the Eastern and Southern coasts of the US, and East Asia, particularly in Japan. In the ocean, there are frequent red tide algae blooms that kill fish and marine mammals and cause respiratory problems in humans and some domestic animals when the blooms reach close to shore. In addition to land runoff, atmospheric anthropogenic fixed nitrogen can enter the open ocean. A study in 2008 found that this could account for around one third of the ocean's external (non-recycled) nitrogen supply and up to three per cent of the annual new marine biological production. It has been been suggested that accumulating reactive nitrogen in the environment may have consequences as serious as putting carbon dioxide in the atmosphere.

Acidification

The oceans are normally a natural carbon sink, absorbing carbon dioxide from the atmosphere. Because the levels of atmospheric carbon dioxide are increasing, the oceans are becoming more acidic. The potential consequences of ocean acidification are not fully understood, but there are concerns that structures made of calcium carbonate may become vulnerable to dissolution, affecting corals and the ability of shellfish to form shells.

A report from NOAA scientists published in the journal Science in May 2008 found that large amounts of relatively acidified water are upwelling to within four miles of the Pacific continental shelf area of North America.

This area is a critical zone where most local marine life lives or is born. While the paper dealt only with areas from Vancouver to northern California, other continental shelf areas may be experiencing similar effects.

Specific Examples

- Chinese and Russian industrial pollution such as phenols and heavy metals in the Amur River have devastated fish stocks and damaged its estuary soil.
- Wabamun Lake in Alberta, Canada, once the best whitefish lake in the area, now has unacceptable levels of heavy metals in its sediment and fish.
- Acute and chronic pollution events have been shown to impact southern California kelp forests, though the intensity of the impact seems to depend on both the nature of the contaminants and duration of exposure.
- Ballast water taken up at sea and released in port is a major source of unwanted exotic marine life. The invasive freshwater zebra mussels, native to the Black, Caspian and Azov seas, were probably transported to the Great Lakes via ballast water from a transoceanic vessel.
- Due to their high position in the food chain and the subsequent accumulation of heavy metals from their diet, mercury levels can be high in larger species such as bluefin and albacore. As a result, in March 2004 the United States FDA issued guidelines recommending that pregnant women, nursing mothers and children limit their intake of tuna and other types of predatory fish.
- Some shellfish and crabs can survive polluted environments, accumulating heavy metals or toxins in their tissues. For example, mitten crabs have a remarkable ability to survive in highly modified aquatic habitats, including polluted waters. The farming and harvesting of such species needs careful management if they are to be used as a food.
- Mining has a poor environmental track record. For example, according to the United States Environmental Protection Agency, mining has contaminated portions of the headwaters of over 40 per cent of

watersheds in the western continental US. Much of this pollution finishes up in the sea.
- Heavy metals enter the environment through oil spills - such as the Prestige oil spill on the Galician coast - or from other natural or anthropogenic sources.

SOLUTIONS

Marine pollution is part of the problem of too much pollution by humans in general. There are only two ways to remedy this: either the human population is reduced, or the ecological footprint left behind by the average human is reduced. If we do not follow the second way, then the first way may be imposed upon us, as world ecosystems falter and cease to support us.

The second way is for us, individually, to consume and pollute less than we do currently. For this there must be social and political will, together with a shift in awareness, so more people respect their environment and are less disposed to abuse it.

At an operational level, regulations, and international government participation is needed. It is often very difficult to regulate marine pollution because pollution spreads over international barriers, thus making regulations hard to create as well as enforce.

Perhaps the most important strategy for reducing marine pollution is education. Most are unaware of the sources, and harmful effects of marine pollution, and therefore little is done to address the situation. In order to inform the population of all the facts, in depth research must be done to provide the full scale of the situation. Then this information must be made public.

As expressed in Daoji and Dag's research, one of the reasons why environmental concern is lacking among the Chinese is because the public awareness is low and therefore should be targeted. Likewise, regulation, based upon such in-depth research should be employed. In California, such regulations have already been put in place to protect Californian coastal waters from agricultural runoff.

This includes the California Water Code, as well as several voluntary programmes. Similarly, in India, several tactics have been employed that help reduce marine pollution, however, they do not significantly target the problem. In Chennai city, India, sewage has been dumped further into open waters. Due to the mass of waste being deposited, open-ocean is best for diluting, and dispersing pollutants, thus making them less harmful to marine ecosystems.

WATER POLLUTION

Water pollution is a major problem in the global context. It has been suggested that it is the leading worldwide cause of deaths and diseases, and that it accounts for the deaths of more than 14,000 people daily. In addition to the acute problems of water pollution in developing countries, industrialized countries continue to struggle with pollution problems as well.

In the most recent national report on water quality in the United States, 45 percent of assessed stream miles, 47 percent of assessed lake acres, and 32 percent of assessed bay and estuarine square miles were classified as polluted.

Water is typically referred to as polluted when it is impaired by anthropogenic contaminants and either does not support a human use, like serving as drinking water, and/or undergoes a marked shift in its ability to support its constituent biotic communities, such as fish. Natural phenomena such as volcanoes, algae blooms, storms, and earthquakes also cause major changes in water quality and the ecological status of water. Water pollution has many causes and characteristics.

WATER POLLUTION CATEGORIES

Surface water and groundwater have often been studied and managed as separate resources, although they are interelated. Sources of surface water pollution are generally grouped into two categories based on their origin.

Point Source Pollution

Point source pollution refers to contaminants that enter a waterway through a discrete conveyance, such as a pipe or ditch. Examples of sources in this category include discharges from a sewage treatment plant, a factory, or a city storm drain. The U.S. Clean Water Act (CWA) defines point source for regulatory enforcement purposes.

Non-point Source Pollution

Non-point source (NPS) pollution refers to diffuse contamination that does not originate from a single discrete source. NPS pollution is often a cumulative effect of small amounts of contaminants gathered from a large area. Nutrient runoff in stormwater from "sheet flow" over an agricultural field or a forest are sometimes cited as examples of NPS pollution. Contaminated stormwater washed off of parking lots, roads and highways, called urban runoff, is sometimes included under the category of NPS pollution. However, this runoff is typically channeled into storm drain systems and discharged through pipes to local surface waters, and is a point source. The CWA definition of point source was amended in 1987 to include municipal storm sewer systems, as well as industrial stormwater, such as from construction sites.

Groundwater Pollution

Interactions between groundwater and surface water are complex. Consequently, groundwater pollution, sometimes referred to as groundwater contamination, is not as easily classified as surface water pollution.

By its very nature, groundwater aquifers are susceptible to contamination from sources that may not directly affect surface water bodies, and the distinction of point vs. non-point source may be irrelevant.

A spill of a chemical contaminant on soil, located away from a surface water body, may not necessarily create point source or non-point source pollution, but non-etheless may contaminate the aquifer below. Analysis of groundwater contamination may focus on soil characteristics and hydrology, as well as the nature of the contaminant itself.

Causes of Water Pollution

The specific contaminants leading to pollution in water include a wide spectrum of chemicals, pathogens, and physical or sensory changes such as elevated temperature and discolouration. While many of the chemicals and substances that are regulated may be naturally occurring (calcium, sodium, iron, manganese, etc.) the concentration is often the key in determining what is a natural component of water, and what is a contaminant. Oxygen-depleting substances may be natural materials, such as plant matter (*e.g.* leaves and grass) as well as man-made chemicals. Other natural and anthropogenic substances may cause turbidity (cloudiness) which blocks light and disrupts plant growth, and clogs the gills of some fish species.

Many of the chemical substances are toxic. Pathogens can produce waterborne diseases in either human or animal hosts. Alteration of water's physical chemistry include acidity (change in pH), electrical conductivity, temperature, and eutrophication. Eutrophication is the fertilization of surface water by nutrients that were previously scarce.

Pathogens

Coliform bacteria are a commonly-used bacterial indicator of water pollution, although not an actual cause of disease. Other microorganisms sometimes found in surface waters which have caused human health problems include:

- *Cryptosporidium parvum*
- *Giardia lamblia*
- *Salmonella*
- *Novovirus* and other viruses
- Parasitic worms (helminths).

High levels of pathogens may result from inadequately treated sewage discharges. This can be caused by a sewage plant designed with less than secondary treatment (more typical in less-developed countries). In developed countries, older cities with aging infrastructure may have leaky sewage collection systems (pipes, pumps, valves), which can cause sanitary sewer overflows. Some cities also have combined sewers, which may discharge untreated sewage during rain storms. Pathogen discharges may also be caused by poorly-managed livestock operations.

Chemical and Other Contaminants

Organic water pollutants include:

- Detergents.
- Disinfection by-products found in chemically disinfected drinking water, such as chloroform.
- Food processing waste, which can include oxygen-demanding substances, fats and grease.
- Insecticides and herbicides, a huge range of organohalides and other chemical compounds.
- Petroleum hydrocarbons, including fuels (gasoline, diesel fuel, jet fuels, and fuel oil) and lubricants (motor oil), and fuel combustion byproducts, from stormwater runoff.
- Tree and brush debris from logging operations.
- Volatile organic compounds (VOCs), such as industrial solvents, from improper storage. Chlorinated solvents, which are dense non-aqueous phase liquids (DNAPLs), may fall to the bottom of reservoirs, since they don't mix well with water and are denser.
- Various chemical compounds found in personal hygiene and cosmetic products.

Inorganic water pollutants include:

- Acidity caused by industrial discharges (especially sulfur dioxide from power plants).
- Ammonia from food processing waste.
- Chemical waste as industrial by-products.
- Fertilizers containing nutrients—nitrates and phosphates—which are found in stormwater runoff from agriculture, as well as commercial and residential use.
- Heavy metals from motor vehicles (via urban stormwater runoff) and acid mine drainage.
- Silt (sediment) in runoff from construction sites, logging, slash and burn practices or land clearing sites.

Macroscopic pollution—large visible items polluting the water—may be termed "floatables" in an urban stormwater context, or marine debris when found on the open seas, and can include such items as:

- Trash (*e.g.* paper, plastic, or food waste) discarded by people on the ground, and that are washed by rainfall into storm drains and eventually discharged into surface waters.
- Nurdles, small ubiquitous waterborne plastic pellets.
- Shipwrecks, large derelict ships.

Thermal Pollution

Thermal pollution is the rise or fall in the temperature of a natural body of water caused by human influence. A common cause of thermal pollution is the use of water as a coolant by power plants and industrial manufacturers.

Elevated water temperatures decreases oxygen levels (which can kill fish) and affects ecosystem composition, such as invasion by new thermophilic species. Urban runoff may also elevate temperature in surface waters.

Thermal pollution can also be caused by the release of very cold water from the base of reservoirs into warmer rivers.

Transport and Chemical Reactions of Water Pollutants

Most water pollutants are eventually carried by rivers into the oceans. In some areas of the world the influence can be traced hundred miles from the mouth by studies using hydrology transport models. Advanced computer models such as SWMM or the DSSAM Model have been used in many locations worldwide to examine the fate of pollutants in aquatic systems.

Indicator filter feeding species such as copepods have also been used to study pollutant fates in the New York Bight, for example. The highest toxin loads are not directly at the mouth of the Hudson River, but 100 kilometers south, since several days are required for incorporation into planktonic tissue. The Hudson discharge flows south along the coast due to coriolis force.

Further south then are areas of oxygen depletion, caused by chemicals using up oxygen and by algae blooms, caused by excess nutrients from algal cell death and decomposition. Fish and shellfish kills have been reported, because toxins climb the food chain after small fish consume copepods, then large fish eat smaller fish, etc. Each successive step up the food chain causes a stepwise concentration of pollutants such as heavy metals (*e.g.* mercury) and persistent organic pollutants such as DDT. This is known as biomagnification, which is occasionally used interchangeably with bioaccumulation.

Large gyres (vortexes) in the oceans trap floating plastic debris. The North Pacific Gyre for example has collected the so-called "Great Pacific Garbage Patch" that is now estimated at 100 times the size of Texas. Many of these long-lasting pieces wind up in the stomachs of marine birds and animals. This results in obstruction of digestive pathways which leads to reduced appetite or even starvation.

Many chemicals undergo reactive decay or chemically change especially over long periods of time in groundwater reservoirs. A noteworthy class of such chemicals is the chlorinated hydrocarbons such as trichloroethylene (used in industrial metal degreasing and electronics manufacturing) and tetrachloroethylene used in the dry cleaning industry (note latest advances in liquid carbon dioxide in dry cleaning that avoids all use of chemicals). Both of these chemicals, which are carcinogens themselves, undergo partial decomposition reactions, leading to new hazardous chemicals (including dichloroethylene and vinyl chloride). Groundwater pollution is much more difficult to abate than surface pollution because groundwater can move great distances through unseen aquifers. Non-porous aquifers such as clays partially purify water of bacteria by simple filtration (adsorption and absorption), dilution, and, in some cases, chemical reactions and biological activity: however, in some cases, the pollutants merely transform to soil contaminants. Groundwater that moves through cracks and caverns is not filtered and can be transported as easily as surface water. In fact, this can be aggravated by the human tendency to use natural sinkholes as dumps in areas of Karst topography.

There are a variety of secondary effects stemming not from the original pollutant, but a derivative condition. An example is silt-bearing surface runoff, which can inhibit the penetration of sunlight through the water column, hampering photosynthesis in aquatic plants.

MEASUREMENT OF WATER POLLUTION

Water pollution may be analyzed through several broad categories of methods: physical, chemical and biological. Most methods involve collection of samples, followed by specialized analytical tests. Some methods may be conducted *in situ,* without sampling, such as temperature. Government agencies and research organizations have published standardized, validated analytical test methods to facilitate the comparability of results from disparate testing events.

Sampling

Sampling of water for physical or chemical testing can be done by several methods, depending on the accuracy needed and the characteristics of the contaminant. Many contamination events are sharply restricted in time, most commonly in association with rain events. For this reason "grab" samples are often inadequate for fully quantifying contaminant levels. Scientists gathering this type of data often employ auto-sampler devices that pump increments of water at either time or discharge intervals. Sampling for biological testing involves collection of plants and/or animals from the surface water body. Depending on the type of assessment, the organisms may be identified for biosurveys (population counts) and returned to the water body, or they may be dissected for bioassays to determine toxicity.

Physical Testing

Common physical tests of water include temperature, solids concentration and turbidity.

Chemical Testing

Water samples may be examined using the principles of analytical chemistry. Many published test methods are available for both organic and inorganic compounds. Frequently-used methods include pH, biochemical oxygen demand (BOD), chemical oxygen demand (COD), nutrients (nitrate and phosphorus compounds), metals (including copper, zinc, cadmium, lead and mercury), oil and grease, total petroleum hydrocarbons (TPH), and pesticides.

Biological Testing

Biological testing involves the use of plant, animal, and/or microbial indicators to monitor the health of an aquatic ecosystem.

Control of Water Pollution

In urban areas, domestic sewage is typically treated by centralized sewage treatment plants. In the U.S., most of these plants are operated by local government agencies. Municipal treatment plants are designed to control conventional pollutants: BOD and suspended solids.

Well-designed and operated systems (*i.e.*, secondary treatment or better) can remove 90 percent or more of these pollutants. Some plants have additional sub-systems to treat nutrients and pathogens. Most municipal plants are not designed to treat toxic pollutants found in industrial wastewater.

Cities with sanitary sewer overflows or combined sewer overflows employ one or more engineering approaches to reduce discharges of untreated sewage, including:

- Utilizing a green infrastructure approach to improve stormwater management capacity throughout the system.
- Repair and replacement of leaking and malfunctioning equipment.
- Increasing overall hydraulic capacity of the sewage collection system (often a very expensive option).

A household or business not served by a municipal treatment plant may have an individual septic tank, which treats the wastewater on site and discharges into the soil. Alternatively, domestic wastewater may be sent to a nearby privately-owned treatment system (*e.g.* in a rural community).

Industrial Wastewater

Some industrial facilities generate ordinary domestic sewage that can be treated by municipal facilities. Industries that generate wastewater with high concentrations of conventional pollutants (*e.g.* oil and grease), toxic pollutants (*e.g.* heavy metals, volatile organic compounds) or other non-conventional pollutants such as ammonia, need specialized treatment systems.

Some of these facilities can install a pre-treatment system to remove the toxic components, and then send the partially-treated wastewater to the municipal system. Industries generating large volumes of wastewater typically operate their own complete on-site treatment systems.

Some industries have been successful at redesigning their manufacturing processes to reduce or eliminate pollutants, through a process called pollution prevention. Heated water generated by power plants or manufacturing plants may be controlled with:

- Cooling ponds, man-made bodies of water designed for cooling by evaporation, convection, and radiation.
- Cooling towers, which transfer waste heat to the atmosphere through evaporation and/or heat transfer.
- Cogeneration, a process where waste heat is recycled for domestic and/or industrial heating purposes.

Agricultural Waste Water

Nonpoint Source Controls

Sediment (loose soil) washed off fields is the largest source of agricultural pollution in the United States. Farmers may utilize erosion controls to reduce runoff flows and retain soil on their fields. Common techniques include contour plowing, crop mulching, crop rotation, planting perennial crops and installing riparian buffers.

Nutrients (nitrogen and phosphorus) are typically applied to farmland as commercial fertilizer; animal manure; or spraying of municipal or industrial wastewater (effluent) or sludge. Nutrients may also enter runoff from crop residues, irrigation water, wildlife, and atmospheric deposition. Farmers can develop and implement nutrient management plans to reduce excess application of nutrients.

To minimize pesticide impacts, farmers may use Integrated Pest Management (IPM) techniques (which can include biological pest control) to maintain control over pests, reduce reliance on chemical pesticides, and protect water quality.

Point Source Wastewater Treatment

Farms with large livestock and poultry operations, such as factory farms, are called *concentrated animal feeding operations* or *confined animal feeding operations* in the U.S. and are being subject to increasing government regulation. Animal slurries are usually treated by containment in lagoons before disposal by spray or trickle application to grassland. Constructed wetlands are sometimes used to facilitate treatment of animal wastes, as are anaerobic lagoons. Some animal slurries are treated by mixing with straw and composted at high temperature to produce a bacteriologically sterile and friable manure for soil improvement.

Construction Site Stormwater

Sediment from construction sites is managed by installation of:

- Erosion controls, such as mulching and hydroseeding, and
- Sediment controls, such as sediment basins and silt fences.

Discharge of toxic chemicals such as motor fuels and concrete washout is prevented by use of:

- Spill prevention and control plans, and
- Specially-designed containers (*e.g.* for concrete washout) and structures such as overflow controls and diversion berms.

Urban Runoff (stormwater)

Effective control of urban runoff involves reducing the velocity and flow of stormwater, as well as reducing pollutant discharges. Local governments use a variety of stormwater management techniques to reduce the effects of urban runoff. These techniques, called best management practices (BMPs) in the U.S., may focus on water quantity control, while others focus on improving water quality, and some perform both functions.

Pollution prevention practices include low impact development techniques, installation of green roofs and improved chemical handling (*e.g.* management of motor fuels and oil, fertilizers and pesticides). Runoff mitigation systems include infiltration basins, bioretention systems, constructed wetlands, retention basins and similar devices.

Thermal pollution from runoff can be controlled by stormwater management facilities that absorb the runoff or direct it into groundwater, such as bioretention systems and infiltration basins. Retention basins tend to be less effective at reducing temperature, as the water may be heated by the sun before being discharged to a receiving stream.

REGULATORY FRAMEWORK

In developed countries, the primary focus of legislation and efforts to curb water pollution for the past several decades was first aimed at point sources.

As many point sources have been effectively regulated—principally factories and sewage treatment plants—greater attention has been placed on controlling municipal and industrial stormwater discharges, and NPS contributions.

United Kingdom

In the UK there are common law rights (civil rights) to protect the passage of water across land unfettered in either quality of quantity. Criminal laws dating back to the 16th century exercised some control over water pollution but it was not until the Rivers (Prevention of pollution) Acts 1951 - 1961 were enacted that any systematic control over water pollution was established.

These laws were strengthened and extended in the Control of Pollution Act 1984 which has since been updated and modified by a series of further acts. It is a criminal offence to either pollute a lake, river, groundwater or the sea or to discharge any liquid into such water bodies without proper authority.

In England and Wales such permission can only be issued by the Environment Agency and in Scotland by SEPA.

United States

In the USA, concern over water pollution resulted in the enactment of state anti-pollution laws in the latter half of the 19th century, and federal legislation enacted in 1899.

The Refuse Act of the federal Rivers and Harbours Act of 1899 prohibits the disposal of any refuse matter from into either the nation's navigable rivers, lakes, streams, and other navigable bodies of water, or any tributary to such waters, unless one has first obtained a permit. The Water Pollution Control Act, passed in 1948, gave authority to the Surgeon General to reduce water pollution. However, this law did not lead to major reductions in pollution.

Growing public awareness and concern for controlling water pollution led Congress to carry out a major re-write of water pollution law in 1972. The Federal Water Pollution Control Act Amendments of 1972, commonly known as the Clean Water Act (CWA), established the basic mechanisms for controlling point source pollution.

The law mandated the United States Environmental Protection Agency (EPA) to publish and enforce wastewater standards for industry and municipal sewage treatment plants. The Act also continued requirements that EPA and states issue water quality standards for surface water bodies. Congress included authorization in the Act for major public financing to build municipal sewage treatment plants. The 1972 CWA, however, did not require similar regulatory standards for non-point sources.

In 1987, Congress expanded the coverage of the CWA with enactment of the Water Quality Act. These amendments defined both municipal and industrial stormwater discharges as point sources and required these facilities to obtain discharge permits. The 1987 law also re-organized the public financing of municipal treatment projects and created a non-point source demonstration grant programme. Further amplification of the CWA included the enactment of the Great Lakes Legacy Act of 2002.

NOISE POLLUTION

The displeasing sounds which lead to the disruption of the balance of human life and natural things which surround him creates a chaotic situation or condition called noise pollution. Noise pollution has harmful effects on the physiological and psychological health of human beings. The noise pollution effects on health include hypertension, annoyance, high stress levels, aggression, hearing loss, tinnitus, sleep disturbance and many such health problems. Apart from the traffic and machines, poor urban planning is also a cause of noise pollution.

NOISE POLLUTION CAUSES

The sources/causes of noise pollution increased or probably originated with the invention of machines. Let us have a look at the various noise pollution causes in detail.

- *Traffic*: It is the biggest source of noise pollution in today's times, especially in urban areas. In the past few years the number of automobile vehicles has increased manifold. Thus, traffic problems created by these vehicles is an important source of noise pollution. The sound produced by the exhaust systems of trucks, autos, buses, motorcycles, etc. cause a lot of noise.
- *Railway Stations*: Locomotive engines of trains are an important source of noise pollution. Besides, the shunting and switching operations in rail yards too cause noise pollution to a great extent. The whistles and horns being used by the railway employees add to the effect and increase noise pollution.
- *Aircrafts*: The national parks and wastelands were earlier considered to be the pollution free zones and people traveled to these places for relaxation. Nowadays, however, the sound created by low flying military aircrafts causes noise pollution in these areas too.
- *Industrial Noise*: The various machines used in industries are also a major cause of noise pollution. The different machines which are responsible for creating noise include the compressors, motors and other machinery. It is therefore recommended to grow trees in the premises of industries, which act as absorbents of noise. The trees grown in the premises of 'Tata Motors' an automobile company based in India is a perfect example of how to reduce noise pollution in the industrial belts. Read more on effects of industrial noise pollution.
- *Construction Equipment*: The different types of machines and equipments used in construction of roads and buildings is also an important cause behind noise pollution. The machinery used for the purpose of construction includes pneumatic hammers, bulldozers, air compressors, dump trucks, loaders, pavement breakers, etc. is also responsible for noise pollution to a great extent.
- *Household Equipment*: The household equipments are amongst the minor contributors to noise pollution. These machines include mixers, vacuum cleaners and other noise creating equipments. The effects of noise pollution caused by these machines shouldn't be neglected. This is because one has to work in close contact with the equipments and the magnitude of sound that one needs to bear is high.
- *Other Causes*: The boilers, plumbing equipment, air conditioners, generators and fans contribute to noise pollution to some extent. Noise created by people in public places too is a major contributor

amongst different sources. Loudspeakers used in public places are responsible for creating noise of a high degree/amplitude.

NOISE POLLUTION SOLUTIONS

The different measures that one could undertake in order to reduce noise pollution are mentioned below. The major noise pollution causes or sources are the automobiles and different types of machines. Regular servicing of automobiles could help in reducing the intensity of sound produced by them. Lubricating the machines used in day-to-day life prove to be helpful in their smooth (noiseless) functioning. Use of loudspeakers should be prohibited in public places. Soundproofing materials need to be used in the house to reduce nose pollution.

SOURCES

The sources of noise pollution vary. In some places noise from construction projects predominates, while in others it is vehicular traffic or noise from airports. Other sources include the noise in occupational settings or even the noise of simultaneous conversations. It also seems from a number of studies that intermittent noise is more of a problem than noise of a similar intensity which is constant.

EFFECTS OF NOISE POLLUTION

Noise pollution affects nearly every aspect of life and probably has damaging physical effects as well. The best-studied and best-defined effect on physical health is the effect of noise on hearing. The research results are clear: loud or sustained noise can damage hearing. The source of the noise is not very important; it can be a pile driver or rock music. What is important is that it can have a lasting impact. Noise pollution also impacts people's sleep. It can result in mood problems and adversely affect job performance.

Several research studies suggest that noise can cause high blood pressure. Others say that psychiatric diseases can be caused by noise. Some of these studies are controversial and are contested by other researchers because so many variables such as age, overall state of health, diet, smoking and drinking habits, socioeconomic factors, and other sources of environmental and social stress must also be taken into account. It is clear, however, that noise, even though a "non-specific stressor", does cause a physical response.

It elicits the same responses as a perceived physical threat would produce: it activates the nervous system, causes the muscles to tense and the heart rate and respiratory rate to increase and prepares the body to fight or to run away. This response-called the "fight or flight" response–underlies all responses to stress. The long-term effects of this kind of stimulus, of being ready to flee or give battle, are not completely understood. Being continuously under stress is something like sitting on the edge of your chair or waiting for the other shoe to drop. Your body isn't quite sure what will happen next or how to respond, and that state constant confusion has been implicated in the development of a number of diseases.

It is also important to remember that people who sense that they have some control over what happens in their lives are impacted less strongly by stressors than those who feel they have no control, and noise is something over which we have very little control.

Noise affects us in another significant way: people exposed to noise feel a greater sense of frustration and annoyance than people whose environment is not as noisy. Annoyance is the expression of the negative feelings experienced when one's activities or the enjoyment of one's surroundings are disrupted. Annoyance can have a major impact on the quality of life and is generally a variable examined when studying the impact of noise. In addition to the other environmental pollutants, noise can affect not only our moods but also our physical well being, and, just like water and air pollution, must be subject to greater study and more stringent controls.

HOW TO LOWER THE POLLUTION

There are some things you can do to help yourself while governments get around to tightening the standards for noise pollution, though:

- You can use air conditioning to allow you to keep windows closed during the noisiest times of the day. This is of course only an interim solution, since air conditioning uses more electricity which raises your energy costs and also requires more power plants which in turn create more air or other forms of pollution. Furthermore, when rooms are closed up indoor air pollution becomes a problem. But it can be a good short-term solution.
- You can buy small noise-canceling devices which sample the frequencies of sound and create other sound waves which in essence collide with the noxious sounds and batter them into other, less disturbing sounds. These devices are relatively new and have not been proven to be fully effective, though.
- You can use other sound-generating devices such as stereo systems, which cover up some of the more disturbing sounds with more pleasurable ones. This a short-term solution, however, since the underlying sounds are still present.
- You can learn some of the techniques described in our section on Stress to lessen the impact unpleasant sounds may have on you. When you are more relaxed in general, big annoyances become little annoyances and may disappear altogether. When you learn to truly relax, you may find that sounds which were once of great concern simply fade away into the background.

- And of course, if none of the other suggestions works, you can become active in your community to work with your local authorities to devise solutions to the problem of noise pollution which may be uniquely suited to you and to the place you live, making life wherever you may be better for everyone.

NOISE EXPOSURE LEVEL AND STANDARDS

The Safety, Health and Welfare at Work revoke and replace the Safety, Health and Welfare at Work (Control of Noise at Work) Regulations 2006 save that paragraph of those Regulations, relating to the entertainment sector, remains in effect until 14 February 2008. These Regulations cover Directive 2003/10/EC of the European Parliament on the minimum health and safety requirements regarding the exposure of employees to the risks arising from physical agents (noise). The regulations apply to all workers and workplaces except crew members in sea or air transport. They are aimed at protecting workers from the potential risk to their hearing caused by exposure to high noise levels or to high noise levels over a long period of time and therefore are aimed at preventing hearing loss and deafness.

The regulations require that the employer must ensure that:

- Noise experienced at work shall be assessed and, when necessary, measured. Formal records of any assessments must be kept in the format set down in the regulations
- The records must be kept for at least 3 years and made available to the employees (or their representatives) and inspectors from the HSA.
- Every employer shall reduce the risks resulting from exposure to noise in the workplace to the lowest level reasonably practicable, taking account of technical progress and the availability of measures to control the noise in particular, at source.
- Where the exposure to noise is likely to exceed 80db(A), the employer shall supply sufficient numbers of suitable and adequate personal ear protectors to the employees, following consultation with the workers concerned regarding suitability and adequacy of the models of hearing protection chosen.
- Where it is not reasonably practicable to reduce the daily personal noise exposure of a worker below 80dB (A), the employer shall make hearing checks available to the worker.
- Penalties be imposed for a breach of the regulations
- Health Surveillance must be provided to all employees, if exposure to noise exceeds, 80dB (A)

As regards standards, the Regulations require an employer to take specific action at certain action values. These are the daily noise exposure level or the peak sound pressure level which, if exceeded, for an employee, action will need to be taken to reduce the risk.

These relate to:

- The levels of exposure to noise of your employees averaged over a working day or week; and
- The maximum noises (peak sound pressure) to which employees are exposed in a working day.

The values involved are:

- *Lower exposure action values*:
  - Daily or weekly exposure of 80 dB;
  - Peak sound pressure of 135 dB;
- *Upper exposure action values*:
  - Daily or weekly exposure of 85 dB;
  - Peak sound pressure of 137 dB.

The daily noise exposure or a peak sound pressure level must not be exceeded for an employee in the workplace. When an employer is determining the exposure limit value they must take account of the attenuation provided by individual hearing protectors which are worn by the employees. The peak sound pressure is the maximum value of the 'C' frequency weighted instantaneous noise pressure.

This can be short bursts of noise energy at any given time. Sound pressure (p) is the average variation in atmospheric pressure caused by the sound. The pressure is continuously varying between positive and negative values, so the average variation is measured by its root mean square. The weekly noise exposure level is the time weighted average of the daily noise exposure levels for a normal week of a five eight hour working days. This may be used by employer to determine the noise exposure to employees if the daily noise exposure varies from one working day to the next.

An employer can use weekly noise exposure level, when there is a constant variation in the daily noise exposure from one working day to the next for the purposes of applying the exposure limit values and the exposure action values, to assess the levels of noise to which his or her employees are exposed, provided that;

- The weekly noise exposure level as shown by adequate monitoring does not exceed the exposure limit value of 87db (A), and
- Appropriate measures are taken in order to reduce the risk associated with these activities to a minimum.

When the level is between 80dB (A) and 85dB (A); the employer must (in addition to reducing noise levels by general action) inform each potentially affected worker and their representatives of the following:

- The results of the measurements taken of the noise and the possible risk to their hearing

- Results of preventative audiometric testing
- What measures are being introduced to reduce the noise levels in the workplace
- The advisability of wearing hearing protection that's individually fitted and where and how to obtain it
- Hearing checks that are available
- Provide any necessary information and where relevant training

When the level is above 85dB (A) the employer must:

- Identify the reasons for the excess noise and put in place a programme to reduce it. This could be either of a technical nature or of organisation of work or both.
- Provide the services of a registered medical practitioner to carry out hearing checks and audiometric testing.
- Put up clearly visible and legible signs indicating that the noise level in the area is likely to exceed 85dB (A), in accordance with the Safety, Health and Welfare at Work.

Workers and their representatives must be informed:

- That the noise level is likely to exceed 85 dBA and of the potential risk of damage to hearing
- About the measurements taken of the noise levels and an explanation of the significance of the results
- About what is being done to reduce the noise levels

An employer is required to reduce the level of noise exposure, or any noise in the workplace to the lowest level reasonably practicable.

This may be done by:

- Reducing the noise at source, for example by enclosing the noise source, relocating noisy machines, by screening affected workers or by providing quiet work areas separated from the noisy area.
- Other methods of reducing the exposure to noise may include reducing the number of employees working in a noisy area, or reducing the length of time they spend in the noisy area or using quieter work methods or processes.

HEARING PROTECTION

- The hearing protection should be worn 100 per cent of the time in noisy environments so as to offer full protective effect
- Roll the plug between your fingers and insert while pulling the outer ear upwards and backwards to straighten the ear canal
- Let the plug expand in the ear for approximately 30 seconds
- Check the plug attenuates noise well, without any leakage
- Wear the plug, preferable with the cord behind the neck
- The earplug should be worn at all times in noisy surrounds

- The plug should be stored before and between usage in a way that protects them from dirt, grease, etc.

What is the best ear defender?

- The one most comfortable for the individual which offers the correct attenuation for the level of risk
- The best ear defender is a cap with a very effective seal
- All ear defenders should be CE marked, type approved against EN352 and accompanied by attenuation data

When wearing hearing protectors:

- 100 per cent Wear Time-Provides predicted protective effect.
- 99 per cent Wear Time-5 minutes carelessness per day reduces the protective effect dramatically
- 90 per cent Wear Time-You can no longer be sure of effective protection

IMPACT OF NOISE CONTROL AND ABATEMENT MEASURES.

Audiometric testing includes baseline audiograms, annual audiograms, training, and follow-up procedures. It is not adequate or safe to believe a HCP is satisfactory merely because a van and technician visit a plant annually and perform audiometric tests that are simply filed into the employees record. This system actually instills a false sense of security since it does not prevent hearing loss; it merely provides computerized data that are often of little use and perhaps not even understood by personnel in some agencies. Such a programme chiefly documents the liability of the agency or university.

Hearing tests are useful because they can detect mild hearing loss before damage is obvious or disabling. Audiograms show the results of hearing tests. They show how intense or loud a sound must be before an employee hears it With annual hearing tests it is possible to trace and document hearing loss, and by so doing to prevent further loss from occurring.

Test results can also be used to indicate whether or the HCP is preventing hearing loss. Test results must be kept for the duration of the employees employment. While exposure to the permissible daily dose 8 hours a day, 40 hours a week, without benefit of hearing protectors for 10,20,30, or even 40 years does not result in material hearing impairment, it is predicted that 10 per cent of employees so exposed will experience a STS of 10 per cent within 10 years.

If hearing tests show an STS has occurred, the employee must be informed within 21 days and counseled about further evaluation. The employee must also be given additional training about use of hearing protectors. In determining whether a STS occurred, allowance may be made for aging (presbycusis). As a rule, revision of a baseline audiogram is not permitted for rehired employees. North Carolina OSHA states, however, that the employer may line out an entry on the OSHA log if the employees period of non-employment is greater than one year and a termination audiogram exists. The rationale for lining the employee out is that he was not exposed to 85 dBA as an 8-hour time-weighted average for that employer in the previous year. The rationale for having a termination audiogram is to make sure the change did not occur between the first test and termination.

Hearing Protectors and Hearing Protector Attenuation

Employers need to enforce use of hearing protectors if employees:

- Receive a dose equal to or greater than the permissible daily dose.
- Receive a dose at or above 50 per cent of the permissible daily dose if they have to wait longer than 6 months to receive their baseline audiograms.
- Have incurred work related standard threshold shifts. Employer also need to make hearing protectors available at no cost to employees routinely exposed to 50 per cent or more of the permissible daily dose.

Basic styles of HPDs in common use include earplugs that are inserted into the ear canal, ear canal caps that seal the canal at or near its rim, and earmuffs that encircle the outer ear. The basic HPDs produce attenuation of noise strictly by passive means without the use of electronic circuitry and include use of materials with high sound-transmission loss properties, liner materials that absorb and dissipate sound, trapped air volumes that provide acoustical impedance, and compliant materials that establish an acoustical seal against the skin. When properly selected for the situation and correctly worn, they provide adequate protection in most environments. However, due to the very nature of the attenuation they provide, they often have an effect on hearing quality and perception.

Special HPDs provide both acceptable attenuation and augmented auditory perception. Technologies which incorporate electronics to achieve such features as noise cancellation, signal transmission, or DC-powered communications capabilities are typically termed active, while those that rely strictly on mechanical means to provide various qualities such as amplitude-sensitive or uniform attenuation are termed passive. Active noise reduction relies on the principle of destructive interference of equal amplitude but exactly out-of-phase sound waves at a given point in space. In the case of HPDs the cancellation is established at the ear. There are 2 types: Those designed solely for hearing protection and those designed for 1-or 2-way communication having the associated boom-or throat-mounted microphone and earphone components, commonly referred to as headsets.

The active and passive designs can provide valuable performance advantages to potentially ameliorate situations such as employees needing less attenuation to hear well, overcoming the spectral distortion typical of conventional HPDs, or providing attenuation that dynamically changes with sound level. When considering specialized HPDs, account must also be made of their increased costs and potential reduced reliability compared with the conventional HPD alternatives. The only certain means for evaluating the effectiveness of personal protectors is to measure periodically the hearing thresholds of the user. If no hearing losses are observed–other than those due to the aging process–the programme my be considered to be successful.

However, a hearing monitoring programme may take years to become meaningful because noise-induced hearing loss usually develops slowly. Protective noise reduction characteristics, or attenuation values, must be used as a short-term guide in the selection of the personal hearing protection equipment for particular noise exposure patterns. An employees deliberate failure to wear hearing protection is not the agencys responsibility and the agency will not be cited for the employees misconduct providing the requirements of the HCP are complied with and that compliance is documented. The adequacy of ear protection depends upon the tightness of the fit and its use by employees. Foam plugs come in one-size that fits everyone and they offer as much protection as rigid plugs.

Employees do, however, need training on how to insert them properly. The correct size of molded or rigid earplugs is vital. Cotton or various swimmers ear plugs provide relatively little protection against dangerous noise. Hearing protectors do not always provide as much attenuation in practice as the manufacturer indicates. OSHA states that ear plugs tested in industrial settings will provide a mean attenuation of about 10 decibels. Given the present state of the art, hearing protectors which offer the greatest attenuation may not reliably reduce exposures to safe levels when an average dose borders on 100 dBA as an 8-hour TWA; *i.e.,* when exposure is more than four times the OSHA permissible daily dose. The combination of plugs and muffs increases attenuation approximately 5 decibels for a total of 15 dBA.

Ear protectors do not make it more difficult to carry on a conversation and hear warnings when noise is above 80 decibels. When noise ranges from 85 to 105 decibels, hearing is actually improved when plugs or muffs are used. This improvement occurs because the levels of both the speech and noise are reduced to where the hearing mechanism is not overloaded and it is relatively free from distortion caused by high levels of sound. When noise is below 80 decibels, earplugs may reduce the intelligibility of face-to-face communication because the speech level may fall below the hearing threshold. There is some change in sound quality when plugs or muffs are used but employees can be expected to quickly adjust to this change.

Employees who are exposed to intermittent noise may not always know when ear protectors are needed. As a rule of thumb, if employees have to raise their voice to be heard by someone less than 2 feet away, they need to use ear protectors. By directing sound into the ear, headphones can produce levels loud enough to damage hearing. Headphone users need to hold their head sets an arms length away. If they can hear the sound, the volume is probably too high.

TRAINING PROGRAMMES AND ACCESS TO INFORMATION AND TRAINING MATERIALS

OSHA requires the employer to:

- Schedule employees who receive 50 per cent or more of the permissible daily noise dose in a training programme at the time of initial assignment and annually thereafter.
- *Make sure that training contains, as a minimum, these elements*:
  - The effects of noise on hearing;
  - The purpose of hearing protectors, the advantages, disadvantages, and attenuation of various types,
  - Instructions on selection, fitting, use and care; and
- The purpose of audiometric testing and an explanation of the test procedures. Training videos can be used but an instructor needs to be immediately available to trainees so that they have the opportunity to obtain interactive answers to their questions.
- Schedule additional training when new exposures are created.
- Make a copy of 29 CFR 1910.95 available to employees (*e.g.*, post a copy in the workplace).
- Provide a copy of 29 CFR 1910.95 to employees exposed to 50 per cent or more of the permissible daily dose should they ask for it.
- Provide copies of training materials to OSHA on request.

ULTRASONICS

Ultrasound is high frequency sound that is inaudible to the human ear. The frequency audibility limit of the human ear is about 20 kilohertz (kHz). The frequency limit normally declines with age. Most of the audible noise associated with ultrasonic sources, such as ultrasonic welders or ultrasonic cleaners, consists of subharmonics of the machines fundamental ultrasonic frequencies. For example, many ultrasonic welders have a fundamental operating frequency of 20 kHz. However, a good deal of noise is also present at 10 kHz, the first subharmonic frequency of the 20 kHz operating frequency, and is therefore audible to most persons. Research indicates that ultrasonic noise has little effect on general health unless there is direct body contact with a radiating ultrasonic source. Reported cases of headache and nausea associated with airborne ultrasonic exposures appear to have been caused by high levels of audible noise from source subharmonics.

NON-OCCUPATIONAL NOISE EXPOSURES

Individual risk factors and other exposures explain the large variation in the development of noise-induced hearing loss (NIHL). Such factors include heredity, cholesterol, blood pressure, vibration induced white finger symptom, painkillers, smoking, and exposure to organic solvents. Hearing loss may occur without noise exposure. While age has a lower significance than previously thought, the large variation in NIHL is mostly explained by individual vulnerability factors which are age dependent. Encouraging employees to take their earplugs home and use them during noisy recreational or other activities will reduce the likelihood of compensation claims by reducing employee hearing loss.

Noisy off-the-job activities include home power tool use, lawn and garden care, commercial air travel, nightclub and sporting events attendance, visits to state fairs, theme parks, etc. Employees may want to test their hearing to see if their off-the-job exposures are causing hearing loss. There are two tests that concerned employees can perform. The first one developed by the American Academy of Otolaryngology is shown in Attachment 4. Another quick test is offered by the Occupational Hearing Services. An employee can dial 1-800-222-EARS weekdays between 9: 00 am and 6: 00 pm to hear a two-minute recorded telephone test The record plays a series of quiet tones. If the employee cant hear all eight of them, he will be advised to get a more complete hearing exam. Before calling, the employee needs to make sure that he is using a good quality telephone and is in a quiet room with the TV and air conditioner off.

RECORDKEEPING

OSHA requires employers to:

- Keep training records for two years that include training dates, contents of training, name and qualifications of the trainer, and names and job classifications of the trainees.
- Keep audiometric test records for the duration of the employees employment that include the employees name and job classification, date of the audiogram, the examiners name, date of the last acoustic or exhaustive calibration of the audiometer, and the employees most recent noise exposure assessment.
- Keep records of the background sound pressure levels in audiometric test rooms for the duration of the employees employment.
- Provide training and test records to employees on request.
- Log standard threshold shifts as illnesses on the OSHA 200 form. According to North Carolina OSHA, if hearing loss is work related (*i.e.,* the employees daily noise dose is equal to or greater than the OSHA permissible daily dose) and there is a 10 decibel or greater standard threshold shift in either ear, the hearing loss is a recordable

illness. Federal OSHA proposed on February 2,1996, that any work-related case resulting in an average shift of 15 decibels or more at 2000, 3000, or 4000 Hertz in one or both ears as measured from the employees original baseline be logged. In the OSHA proposal, the hearing test may be adjusted for aging and the recorded case may be removed if a retest performed within 30 days does not confirm the original shift. OSHA states a presumption of work-relatedness is used for hearing loss occurring to employees who are exposed to noise levels above 85 dBA as an 8-hour time-weighted average and proposes that a shift of 10 dB be used as the action level.

7

Health Impact of Nanotechnology

The health impact of nanotechnology are the possible effects that the use of nanotechnological materials and devices will have on human health. As nanotechnology is an emerging field, there is great debate regarding to what extent nanotechnology will benefit or pose risks for human health. Nanotechnology's health impact can be split into two aspects: the potential for nanotechnological innovations to have medical applications to cure disease, and the potential health hazards posed by exposure to nanomaterials.

Nanotoxicology

The extremely small size of nanomaterials also means that they are much more readily taken up by the human body than larger sized particles. How these nanoparticles behave inside the body is one of the issues that needs to be resolved. The behaviour of nanoparticles is a function of their size, shape and surface reactivity with the surrounding tissue.

They could cause overload on phagocytes, cells that ingest and destroy foreign matter, thereby triggering stress reactions that lead to inflammation and weaken the body's defence against other pathogens. Apart from what happens if non-degradable or slowly degradable nanoparticles accumulate in organs, another concern is their potential interaction with biological processes inside the body: because of their large surface, nanoparticles on exposure to tissue and fluids will immediately adsorb onto their surface some of the macromolecules they encounter.

This may, for instance, affect the regulatory mechanisms of enzymes and other proteins.

Other properties of nanomaterials that influence toxicity include: chemical composition, shape, surface structure, surface charge, aggregation and solubility, and the presence or absence of functional groups of other chemicals. The large number of variables influencing toxicity means that it is difficult to generalise about health risks associated with exposure to nanomaterials – each new nanomaterial must be assessed individually and all material properties must be taken into account.

California

In October 2008, the Department of Toxic Substances Control (DTSC), within the California Environmental Protection Agency, announced its intent to request information regarding analytical test methods, fate and transport in the environment, and other relevant information from manufacturers of carbon nanotubes. These sections were added as a result of the adoption of Assembly Bill AB 289 (2006). They are intended to make information on the fate and transport, detection and analysis, and other information on chemicals more available. The law places the responsibility to provide this information to the Department on those who manufacture or import the chemicals.

On January 22, 2009, a formal information request letter was sent to manufacturers who produce or import carbon nanotubes in California, or who may export carbon nanotubes into the State.

This letter constitutes the first formal implementation of the authorities placed into statute by AB 289 and is directed to manufacturers of carbon nanotubes, both industry and academia within the State, and to manufacturers outside California who export carbon nanotubes to California. This request for information must be met by the manufacturers within one year. DTSC is waiting for the upcoming January 22, 2010 deadline for responses to the data call-in. The California Nano Industry Network and DTSC hosted a full-day symposium on November 16, 2009 in Sacramento, CA. This symposium provided an opportunity to hear from nanotechnology industry experts and discuss future regulatory considerations in California. DTSC is expanding the Specific Chemical Information Call-in to members of the nanometal oxides.

Nanomedicine

Nanomedicine is the medical application of nanotechnology. The approaches to nanomedicine range from the medical use of nanomaterials, to nanoelectronic biosensors, and even possible future applications of molecular nanotechnology. Current problems for nanomedicine involve understanding the issues related to toxicity and environmental impact of nanoscale materials.

Nanomedicine research is directly funded, with the US National Institutes of Health in 2005 funding a five-year plan to set up four nanomedicine centres. In April 2006, the journal Nature Materials estimated that 130 nanotech-based drugs and delivery systems were being developed worldwide.

Nanomedicine seeks to deliver a set of research tools and clinical devices in the near future. The National Nanotechnology Initiative expects new commercial applications in the pharmaceutical industry that may include advanced drug delivery systems, new therapies, and *in vivo* imaging. Neuro-electronic interfaces and other nanoelectronics-based sensors are another active goal of research. Further down the line, the speculative field of molecular nanotechnology believes that cell repair machines could revolutionize medicine and the medical field. Nanomedicine is a large industry, with nanomedicine sales reaching $6.8 billion in 2004. With over 200 companies and 38 products worldwide, a minimum of $3.8 billion in nanotechnology R and D is being invested every year. As the nanomedicine industry continues to grow, it is expected to have a significant impact on the economy.

Curing Diseases

Currently, nanotech gene therapy has been able to kill ovarian cancer in mice while avoiding the side effects of *cisplatin* and *paclitaxel*; it is speculated that this technology could save 15000 women in the United States each year if the treatment proves effective and safe in humans. Research on nanoelectronics-based cancer diagnostics could lead to tests that can be done in pharmacies. The results promise to be highly accurate and the product promises to be inexpensive. They could take a very small amount of blood and detect cancer anywhere in the body in about five minutes, with a sensitivity that is a thousand times better than in a conventional laboratory test. These devices that are built with nanowires to detect cancer proteins; each nanowire detector is primed to be sensitive to a different cancer marker. The biggest advantage of the nanowire detectors is that they could test for anywhere from ten to one hundred similar medical conditions without adding cost to the testing device.

IMPACT OF MOLECULAR NANOTECHNOLOGY

Molecular nanotechnology is a speculative subfield of nanotechnology regarding the possibility of engineering molecular assemblers, machines which could re-order matter at a molecular or atomic scale. Regarding the risks from molecular manufacturing, an often cited worst-case scenario is "grey goo", a hypothetical substance into which the surface of the earth might be transformed by self-replicating nanobots running amok. This concept has been analysed by Freitas in "Some Limits to Global Ecophagy by Biovorous Nanoreplicators, with Public Policy Recommendations" With the advent of nan-biotech, a different scenario called green goo has been forwarded. Here, the malignant substance is not nanobots but rather self-replicating organisms engineered through nanotechnology.

According to the Centre for Responsible Nanotechnology:

Molecular manufacturing allows the cheap creation of incredibly powerful devices and products. How many of these products will we want? What environmental damage will they do? The range of possible damage is vast, from personal low-flying supersonic aircraft injuring large numbers of animals to collection of solar energy on a sufficiently large scale to modify the planet's albedo and directly affect the environment. Stronger materials will allow the creation of much larger machines, capable of excavating or otherwise destroying large areas of the planet at a greatly accelerated pace.

It is too early to tell whether there will be economic incentive to do this. However, given the large number of activities and purposes that would damage the environment if taken to extremes, and the ease of taking them to extremes with molecular manufacturing, it seems likely that this problem is worth worrying about. Some forms of damage can result from an aggregate of individual actions, each almost harmless by itself. Such damage is quite hard to prevent by persuasion, and laws frequently don't work either; centralized restriction on the technology itself may be a necessary part of the solution. Finally, the extreme compactness of nanomanufactured machinery will tempt the use of very small products, which can easily turn into nano-litter that will be hard to clean up and may cause health problems. The site list numerous other risks and benefits.

SOCIETAL IMPACT

Beyond the toxicity risks to human health and the environment which are associated with first-generation nanomaterials, nanotechnology has broader societal impact and poses broader social challenges. Social scientists have suggested that nanotechnology's social issues should be understood and assessed not simply as "downstream" risks or impacts. Rather, the challenges should be factored into "upstream" research and decision making in order to ensure technology development that meets social objectives. Many social scientists and organizations in civil society suggest that technology assessment and governance should also involve public participation.

Societal risks from the use of nanotechnology have also been raised. On the instrumental level, these include the possibility of military applications of nanotechnology (for instance, as in implants and other means for soldier enhancement like those being developed at the Institute for Soldier Nanotechnologies at MIT) as well as enhanced surveillance capabilities through nano-sensors.

Potential Benefits and Risks for Developing Countries

Nanotechnologies may provide new solutions for the millions of people in developing countries who lack access to basic services, such as safe water, reliable energy, health care, and education. The United Nations has set Millennium Development Goals for meeting these needs. The 2004 UN Task Force on Science, Technology and Innovation noted that some of the advantages of nanotechnology include production using little labour, land, or maintenance, high productivity, low cost, and modest requirements for materials and energy.

Potential opportunities of nanotechnologies to help address critical international development priorities include improved water purification systems, energy systems, medicine and pharmaceuticals, food production and nutrition, and information and communications technologies. Nanotechnologies are already incorporated in products that are on the market. Other nanotechnologies are still in the research phase, while others are concepts that are years or decades away from development.

Protection of the environment, human health and worker safety in developing countries often suffers from a combination of factors that can include but are not limited to lack of robust environmental, human health, and worker safety regulations; poorly or unenforced regulation which is linked to a lack of physical (*e.g.*, equipment) and human capacity (*i.e.*, properly trained regulatory staff). Often, these nations require assistance, particularly financial assistance, to develop the scientific and institutional capacity to adequately assess and manage risks, including the necessary infrastructure such as laboratories and technology for detection.

However, concerns are frequently raised that the claimed benefits of nanotechnology will not be evenly distributed, and that any benefits (including technical and/or economic) associated with nanotechnology will only reach affluent nations. The majority of nanotechnology research and development-and patents for nanomaterials and products-is concentrated in developed countries (including the United States, Japan, Germany, Canada and France). In addition, most patents related to nanotechnology are concentrated amongst few multinational corporations, including IBM, Micron Technologies, Advanced Micro Devices and Intel. This has led to fears that it will be unlikely that developing countries will have access to the infrastructure, funding and human resources required to support nanotechnology research and development, and that this is likely to exacerbate such inequalities. Producers in developing countries could also be disadvantaged by the replacement of natural products (including rubber, cotton, coffee and tea) by developments in nanotechnology. These natural products are significant export crops for developing countries, and many farmers' livelihoods depend on them. It has been argued that their substitution with industrial nano-products could negatively affect the economies of developing countries, that have traditionally relied on these export crops.

Effects on Labourers

Ray Kurzweil has speculated in *The Singularity is Near* that people who work in unskilled labour jobs for a livelihood may become the first human workers to be displaced by the constant use of nanotechnology in the workplace, noting that layoffs often affect the jobs based around the lowest technology level before attacking jobs with the highest technology level possible. It has been noted that every major economic era has stimulated a global revolution both in the kinds of jobs that are available to people and the kind of training they need to achieve these jobs, and there is concern that the world's educational systems have lagged behind in preparing students for the "Nanotech Age".

It has also been speculated that nanotechnology may give rise to nanofactories which may have superior capabilities to conventional factories due to their small carbon and physical footprint on the global and regional environment. The miniaturization and transformation of the multi-acre conventional factory into the nanofactory may not interfere with their ability to deliver a high quality product; the product may be of even greater quality due to the lack of human errors in the production stages. Nanofactory systems may use precise atomic precisioning and contribute to making superior quality products that the "bulk chemistry" method used in 20th century and early 21st currently cannot produce. These advances might shift the computerized workforce in an even more complex direction, requiring skills in genetics, nanotechnology, and robotics.

SOCIETAL IMPACT OF NANOTECHNOLOGY

The societal impact of nanotechnology are the potential benefits and challenges that the introduction of novel nanotechnological devices and materials may hold for society and human interaction. The term is sometimes expanded to also include nanotechnology's health and environmental impact, but this article will only consider the social and political impact of nanotechnology. As nanotechnology is an emerging field and most of its applications are still speculative, there is much debate about what positive and negative effects that nanotechnology might have.

Beyond the toxicity risks to human health and the environment which are associated with first-generation nanomaterials, nanotechnology has broader societal implications and poses broader social challenges. Social scientists have suggested that nanotechnology's social issues should be understood and assessed not simply as "downstream" risks or impacts. Rather, the challenges should be factored into "upstream" research and decision making in order to ensure technology development that meets social objectives

Many social scientists and organizations in civil society suggest that technology assessment and governance should also involve public participation Some observers suggest that nanotechnology will build incrementally, as did the 18-19th century industrial revolution, until it gathers pace to drive a nanotechnological revolution that will radically reshape our economies, our labour markets, international trade, international relations, social structures, civil liberties, our relationship with the natural world and even what we understand to be human. Others suggest that it may be more accurate to describe change driven by nanotechnology as a "technological tsunami". Just like a tsunami, analysts warn that rapid nanotechnology-driven change will necessarily have profound disruptive impacts. As the APEC Centre for Technology Foresight observes: If nanotechnology is going to revolutionize manufacturing, health care, energy supply, communications and probably defence, then it will transform labour and the workplace, the medical system, the transportation and power infrastructures and the military. None of these latter will be changed without significant social disruption. Those concerned with the negative impact of nanotechnology suggest that it will simply exacerbate problems stemming from existing socio-economic inequity and unequal distributions of power, creating greater inequities between rich and poor through an inevitable nano-divide (the gap between those who control the new nanotechnologies and those whose products, services or labour are displaced by them). Analysts suggest the possibility that nanotechnology has the potential to destabilize international relations through a nano arms race and the increased potential for bioweaponry; thus, providing the tools for ubiquitous surveillance with significant implications for civil liberties. Also, many critics believe it might break down the barriers between life and non-life through nanobiotechnology, redefining even what it means to be human. Nanoethicists posit that such a transformative technology could exacerbate the divisions of rich and poor – the so-called "nano divide." However nanotechnology makes the production of technology, *e.g.* computers, cellular phones, health technology etcetera, cheaper and therefore accessible to the poor.

In fact, many of the most enthusiastic proponents of nanotechnology, such as transhumanists, see the nascent science as a mechanism to changing human nature itself – going beyond curing disease and enhancing human characteristics. Discussions on nanoethics have been hosted by the federal government, especially in the context of "converging technologies" – a catch-phrase used to refer to nano, biotech, information technology, and cognitive science.

Possible Military Applications

Societal risks from the use of nanotechnology have also been raised. On the instrumental level, these include the possibility of military applications of nanotechnology (for instance, as in implants and other means for soldier enhancement like those being developed at the Institute for Soldier Nanotechnologies at MIT) as well as enhanced surveillance capabilities through nano-sensors. There is also the possibility of nanotechnology being used to develop chemical weapons and because they will be able to develop the chemicals from the atom scale up, critics fear that chemical weapons developed from nano particles will be more dangerous than present chemical weapons.

Intellectual Property Issues

On the structural level, critics of nanotechnology point to a new world of ownership and corporate control opened up by nanotechnology. The claim is that, just as biotechnology's ability to manipulate genes went hand in hand with the patenting of life, so too nanotechnology's ability to manipulate molecules has led to the patenting of matter. The last few years has seen a gold rush to claim patents at the nanoscale. Over 800 nano-related patents were granted in 2003, and the numbers are increasing year to year. Corporations are already taking out broad-ranging patents on nanoscale discoveries and inventions. For example, two corporations, NEC and IBM, hold the basic patents on carbon nanotubes, one of the current cornerstones of nanotechnology. Carbon nanotubes have a wide range of uses, and look set to become crucial to several industries from electronics and computers, to strengthened materials to drug delivery and diagnostics. Carbon nanotubes are poised to become a major traded commodity with the potential to replace major conventional raw materials. However, as their use expands, anyone seeking to (legally) manufacture or sell carbon nanotubes, no matter what the application, must first buy a license from NEC or IBM.

The United States' essential facilities doctrine may be of importance as well as other anti-trust laws.

POTENTIAL SOCIAL IMPACTS

Benefits

Nanotechnology will let us continue the historical trends in manufacturing right up to the fundamental limits imposed by physical law. It will let us make remarkably powerful molecular computers. It will let us make materials over fifty times lighter than steel or aluminium alloy but with the same strength. We'll be able to make jets, rockets, cars or even chairs that, by today's standards, would be remarkably light, strong, and inexpensive. Molecular surgical tools, guided by molecular computers and injected into the blood stream could find and destroy cancer cells or invading bacteria, unclog arteries, or provide oxygen when the circulation is impaired.

Nanotechnology will replace our entire manufacturing base with a new, radically more precise, radically less expensive, and radically more flexible way of making products. The aim is not simply to replace today's computer chip making plants, but also to replace the assembly lines for cars, televisions, telephones, books, surgical tools, missiles, bookcases, airplanes, tractors, and all the rest. The objective is a pervasive change in manufacturing, a change that will leave virtually no product untouched. Economic progress and military readiness in the 21st Century will depend fundamentally on maintaining a competitive position in nanotechnology.

Despite the current early developmental status of nanotechnology and molecular nanotechnology, much concern surrounds MNT's anticipated impact on economics and on law. Whatever the exact effects, MNT, if achieved, would tend to reduce the scarcity of manufactured goods and make many more goods (such as food and health aids) manufacturable.

It is generally considered that future citizens of a molecular-nanotechnological society would still need money, in the form of unforgeable digital cash or physical specie (in special circumstances). They might use such money to buy goods and services that are unique, or limited within the solar system. These might include: matter, energy, information, real estate, design services, entertainment services, legal services, fame, political power, or the attention of other people to ones political/religious/philosophical message. Furthermore, futurists must consider war, even between prosperous states, and non-economic goals.

If MNT were realized, some resources would remain limited, because unique physical objects are limited (a plot of land in the real Jerusalem, mining rights to the larger near-earth asteroids) or because they depend on the goodwill of a particular person (the love of a famous person, a live audience in a musical concert). Demand will always exceed supply for some things, and a political economy may continue to exist in any case. Whether the interest in these limited resources would diminish with the advent of virtual reality, where they could be easily substituted, is yet unclear. One reason why it might not is a hypothetical preference for "the real thing", although such an opinion could easily be mollified if virtual reality were to develop to a certain level of quality.

MNT should make possible nanomedical capabilities able to cure any medical condition not already cured by advances in other areas. Good health would be common, and poor health of any form would be as rare as smallpox and scurvy are today. Even cryonics would be feasible, as cryopreserved tissue could be fully repaired.

Risks

Molecular nanotechnology is one of the technologies that some analysts believe could lead to a Technological Singularity. Some feel that molecular nanotechnology would have daunting risks. It conceivably could enable cheaper and more destructive conventional weapons.

Also, molecular nanotechnology might permit weapons of mass destruction that could self-replicate, as viruses and cancer cells do when attacking the human body. Commentators generally agree that, in the event molecular nanotechnology were developed, mankind should permit self-replication only under very controlled or "inherently safe" conditions.

A fear exists that nanomechanical robots, if achieved, and if designed to self-replicate using naturally occurring materials (a difficult task), could consume the entire planet in their hunger for raw materials, or simply crowd out natural life, out-competing it for energy (as happened historically when blue-green algae appeared and outcompeted earlier life forms). Some commentators have referred to this situation as the "grey goo" or "ecophagy" scenario. K. Eric Drexler considers an accidental "grey goo" scenario extremely unlikely and says so in later editions of *Engines of Creation*.

In light of this perception of potential danger, the Foresight Institute (founded by K. Eric Drexler to prepare for the arrival of future technologies) has drafted a set of guidelines for the ethical development of nanotechnology.

These include the banning of free-foraging self-replicating pseudo-organisms on the Earth's surface, at least, and possibly in other places.

ENVIRONMENTAL BENEFITS OF NANOTECHNOLOGY

ENERGY

Nanotechnology could potentially have a great impact on clean energy production. Research is underway to use nanomaterials for purposes including more efficient solar cells, practical fuel cells, and environmentally friendly batteries. The most advanced nanotechnology projects related to energy are: storage, conversion, manufacturing improvements by reducing materials and process rates, energy saving (by better thermal insulation for example), and enhanced renewable energy sources.

Current commercially available solar cells have low efficiencies of 15-20 per cent. Research is ongoing to use nanowires and other nanostructured materials with the hope of to create cheaper and more efficient solar cells than are possible with conventional planar silicon solar cells. It is believed that these nanoelectronics-based devices will enable more efficient solar cells, and would have a great effect on satisfying global energy needs.

Another example for an environmentally friendly form of energy is the use of fuel cells powered by hydrogen. Probably the most prominent nanostructured material in fuel cells is the catalyst consisting of carbon supported noble metal particles with diameters of 1-5 nm. Suitable materials for hydrogen storage contain a large number of small nanosized pores.

Nanotechnology may also find applications in batteries. Because of the relatively low energy density of conventional batteries the operating time is limited and a replacement or recharging is needed, and the huge number of spent batteries represent a disposal problem. The use of nanomaterials may enable batteries with higher energy content or supercapacitors with a higher rate of recharging, which could be helpful for the battery disposal problem.

Water Filtration and Remediation

A strong influence of nanochemistry on waste-water treatment, air purification and energy storage devices is to be expected. Mechanical or chemical methods can bc used for effective filtration techniques. One class of filtration techniques is based on the use of membranes with suitable hole sizes, whereby the liquid is pressed through the membrane. Nanoporous membranes are suitable for a mechanical filtration with extremely small pores smaller than 10 nm ("nanofiltration") and may be composed of nanotubes. Nanofiltration is mainly used for the removal of ions or the separation of different fluids. Magnetic nanoparticles offer an effective and reliable method to remove heavy metal contaminants from waste water by making use of magnetic separation techniques. Using nanoscale particles increases the efficiency to absorb the contaminants and is comparatively inexpensive compared to traditional precipitation and filtration methods. Some water-treatment devices incorporating nanotechnology are already on the market, with more in development. Low-cost nanostructured separation membranes methods have been shown to be effective in producing potable water in a recent study. Nanoscale iron particles have also shown potential as a detoxifying agent for cleaning environmental contaminents from brownfield sites.

HUMAN AND ENVIRONMENTAL HEALTH AND SAFETY

Studies of the health impact of airborne particles generally shown that for toxic materials, smaller particles are more toxic. This is due in part to the fact that, given the same mass per volume, the dose in terms of particle numbers increases as particle size decreases. Based upon available data, it has been argued that current risk assessment methodologies are not suited to the hazards associated with nanoparticles; in particular, existing toxicological and eco-toxicological methods are not up to the task; exposure evaluation (dose) needs to be expressed as quantity of nanoparticles and/or surface area rather than simply mass; equipment for routine detecting and measuring nanoparticles in air, water, or soil is inadequate; and very little is known about the physiological responses to nanoparticles.

Regulatory bodies in the U.S. as well as in the EU have concluded that nanoparticles form the potential for an entirely new risk and that it is necessary to carry out an extensive analysis of the risk. The challenge for regulators is whether a matrix can be developed which would identify nanoparticles and more complex nanoformulations which are likely to have special toxicological properties or whether it is more reasonable for each particle or formulation to be tested separately.

The International Council on Nanotechnology maintains a database and Virtual Journal of scientific papers on environmental, health and safety research on nanoparticles. The database currently has over 2000 entries indexed by particle type, exposure pathway and other criteria. The Project on Emerging Nanotechnologies (PEN) currently lists 807 products that manufacturers have voluntarily identified that use nanotechnology. No labelling is required by the FDA so that number could be significantly higher. "The use of nanotechnology in consumer products and industrial applications is growing rapidly, with the products listed in the PEN inventory showing just the tip of the iceberg" according to PEN Project Director David Rejeski. A list of those products that have been voluntarily disclosed by their manufacturers is located here.

The Material Safety Data Sheet that must be issued for certain materials often does not differentiate between bulk and nanoscale size of the material in question and even when it does these MSDS are advisory only.

Democratic Governance

Many argue that government has a responsibility to provide opportunities for the public to be involved in the development of new forms of science and technology. Community engagement can be achieved through various means or mechanisms. An online journal article identifies traditional approaches such as referenda, consultation documents, and advisory committees that include community members and other stakeholders. Other conventional approaches include public meetings and "closed" dialog with stakeholders. More contemporary engagement processes that have been employed to include community members in decisions about nanotechnology include citizens' juries and consensus conferences. Leach and Scoones argue that since that "most debates about science and technology options involve uncertainty, and often ignorance, public debate about regulatory regimes is essential."

It has been argued that limited nanotechnology labelling and regulation may exacerbate potential human and environmental health and safety issues associated with nanotechnology, and that the development of comprehensive regulation of nanotechnology will be vital to ensure that the potential risks associated with the research and commercial application of nanotechnology do not overshadow its potential benefits. Regulation may also be required to meet community expectations about responsible development of nanotechnology, as well as ensuring that public interests are included in shaping the development of nanotechnology.

Community education, engagement and consultation tend to occur "downstream": once there is at least a moderate level of awareness, and often during the process of disseminating and adapting technologies. "Upstream" engagement, by contrast, occurs much earlier in the innovation cycle and involves: "dialogue and debate about future technology options and pathways, bringing the often expert-led approaches to horizon scanning, technology foresight and scenario planning to involve a wider range of perspectives and inputs." Daniel Sarewitz Director of Arizona State University's Consortium on Science, Policy and Outcomes, argues that "by the time new devices reach the stage of commercialization and regulation, it is usually too late to alter them to correct problems." The stance that the research, development and use of nanotechnology should be subject to control by the public sector is sometimes referred to as nanosocialism.

Newness

The question of whether nanotechnology represents something 'new' must be answered to decide how best nanotechnology should be regulated. The Royal Society recommended that the UK government assess chemicals in the form of nanoparticles or nanotubes as new substances. Subsequent to this, in 2007 a coalition of over forty groups called for nanomaterials to be classified as new substances, and regulated as such. Despite these recommendations, chemicals comprising nanoparticles that have previously been subject to assessment and regulation may be exempt from regulation, regardless of the potential for different risks and impacts. In contrast, nanomaterials are often recognized as 'new' from the perspective of intellectual property rights (IPRs), and as such are commercially protected via patenting laws.

New Regulatory Framework, or Adapt Existing Arrangements?

There is significant debate about who is responsible for the regulation of nanotechnology. While some non-nanotechnology specific regulatory agencies currently cover some products and processes (to varying degrees) – by "bolting on" nanotechnology to existing regulations – there are clear gaps in these regimes. This enables some nanotechnology applications to figuratively "slip through the cracks" without being covered by any regulations. An example of this has occurred in the US, and involves nanoparticles of titanium dioxide (TIO_2) for use in sunscreen where they create a clearer cosmetic appearance. In this case, the US Food and Drug Administration (FDA) reviewed the immediate health effects of exposure to nanoparticles of titanium dioxide (TIO_2) for consumers. However, they did not review its impacts for aquatic ecosystems when the sunscreen rubs off, nor did the EPA, or any other agency.

Similarly the Australian equivalent of the FDA, the Therapeutic Goods Administration (TGA) approved the use of nanoparticles in sunscreens (without the requirement for package labelling) after a thorough review of the literature, on the basis that although nanoparticles of TIO_2 and zinc oxide (ZNO) in sunscreens do produce free radicals and oxidative DNA damage *in vitro*, such particles were unlikely to pass the dead outer cells of the stratum corneum of human skin; a finding which some academics have argued seemed not to apply the precautionary principle in relation to prolonged use on children with cut skin, the elderly with thin skin, people with diseased skin or use over flexural creases.

Doubts over the TGA's decision were raised with publication of a paper showing that the uncoated anatase form of TIO_2 used in some Australian sunscreens caused a photocatalytic reaction that degraded the surface of newly installed prepainted steel roofs in places where they came in contact with the sunscreen coated hands of workmen. Such gaps in regulation are likely to continue alongside the development and commercialization of increasingly complex second and third generation nanotechnologies. Nanomedicines are just beginning to enter drug regulatory processes, but within a few decades could comprise a dominant group within the class of innovative pharmaceuticals, the current thinking of government safety and cost-effectiveness regulators appearing to be that these products give rise to few if any nano-specific issues. Some academics (such as Thomas Alured Faunce) have challenged that proposition and suggest that nanomedicines may create unique or heightened policy challenges for government systems of cost-effectiveness as well as safety regulation. There are also significant public good aspects to the regulation of nanotechnology, particularly with regard to ensuring that industry involvement in standard-setting does not become a means of reducing competition and that nanotechnology policy and regulation encourages new models of safe drug discovery and development more systematically targeted at the global burden of disease.

The self-regulation attempts may well fail, due to the inherent conflict of interest in asking any organization to police itself. If the public becomes aware of this failure, an external, independent organization is often given the duty of policing them, sometimes with highly punitive measures taken against the organization. The Food and Drug Administration notes that it only regulates on the basis of voluntary claims made by the product manufacturer. If no claims are made by a manufacturer, then the FDA may be unaware of nanotechnology being employed.

Yet regulations worldwide still fail to distinguish between materials in their nanoscale and bulk form. This means that nanomaterials remain effectively unregulated; there is no regulatory requirement for nanomaterials to face new health and safety testing or environmental impact assessment prior to their use in commercial products, if these materials have already been approved in bulk form. The health risks of nanomaterials are of particular concern for workers who may face occupational exposure to nanomaterials at higher levels, and on a more routine basis, than the general public.

International Law

There is no international regulation of nanoproducts or the underlying nanotechnology. Nor are there any internationally agreed definitions or terminology for nanotechnology, no internationally agreed protocols for toxicity testing of nanoparticles, and no standardized protocols for evaluating the environmental impacts of nanoparticles. Since products that are produced using nanotechnologies will likely enter international trade, it is argued that it will be necessary to harmonize nanotechnology standards across national borders. There is concern that some countries, most notably developing countries, will be excluded from international standards negotiations. The Institute for Food and Agricultural Standards notes that "developing countries should have a say in international nanotechnology standards development, even if they lack capacity to enforce the standards".

HEALTH AND SAFETY IMPACT FROM NANOPARTICLES

The presence of nanomaterials (materials that contain nanoparticles) is not in itself a threat. It is only certain aspects that can make them risky, in particular their mobility and their increased reactivity. Only if certain properties of certain nanoparticles were harmful to living beings or the environment would we be faced with a genuine hazard. In this case it can be called nanopollution.

In addressing the health and environmental impact of nanomaterials we need to differentiate between two types of nanostructures:

- Nanocomposites, nanostructured surfaces and nanocomponents (electronic, optical, sensors etc.), where nanoscale particles are incorporated into a substance, material or device ("fixed" nano-particles); and
- "Free" nanoparticles, where at some stage in production or use individual nanoparticles of a substance are present. These free nanoparticles could be nanoscale species of elements, or simple compounds, but also complex compounds where for instance a nanoparticle of a particular element is coated with another substance ("coated" nanoparticle or "core-shell" nanoparticle).

There seems to be consensus that, although one should be aware of materials containing fixed nanoparticles, the immediate concern is with free nanoparticles. Nanoparticles are very different from their everyday counterparts, so their adverse effects cannot be derived from the known toxicity of the macro-sized material. This poses significant issues for addressing the health and environmental impact of free nanoparticles.

To complicate things further, in talking about nanoparticles it is significant that a powder or liquid containing nanoparticles almost never be monodisperse, but contain instead a range of particle sizes. This complicates the experimental analysis as larger nanoparticles might have different properties from smaller ones. Also, nanoparticles show a tendency to aggregate, and such aggregates often behave differently from individual nanoparticles.

The lethal dose over six months for lab rats, of different kinds of nanoparticles are often characterized by a Skov Kjaer index, named after the scientist Kasper Skov Kjaer. The National Institute for Occupational Safety and Health has conducted initial research on how nanoparticles interact with the body's systems and how workers might be exposed to nano-sized particles in the manufacturing or industrial use of nanomaterials. NIOSH currently offers interim guidelines for working with nanomaterials consistent with the best scientific knowledge.

In "The Consumer Product Safety Commission and Nanotechnology," E. Marla Felcher suggests that the Consumer Product Safety Commission, which is charged with protecting the public against unreasonable risks of injury or death associated with consumer products, is ill-equipped to oversee the safety of complex, high-tech products made using nanotechnology.

Longer-term concerns centre on the impact that new technologies will have for society at large, and whether these could possibly lead to either a post-scarcity economy, or alternatively exacerbate the wealth gap between developed and developing nations. The effects of nanotechnology on the society as a whole, on human health and the environment, on trade, on security, on food systems and even on the definition of "human", have not been characterized or politicized.

Health Issues

The health impact of nanotechnology are the possible effects that the use of nanotechnological materials and devices will have on human health. As nanotechnology is an emerging field, there is great debate regarding to what extent nanotechnology will benefit or pose risks for human health. Nanotechnology's health impact can be split into two aspects: the potential for nanotechnological innovations to have medical applications to cure disease, and the potential health hazards posed by exposure to nanomaterials.

Nanotoxicology is the field which studies potential health risks of nanomaterials. The extremely small size of nanomaterials means that they are much more readily taken up by the human body than larger sized particles. How these nanoparticles behave inside the organism is one of the significant issues that needs to be resolved.

The behaviour of nanoparticles is a function of their size, shape and surface reactivity with the surrounding tissue. Apart from what happens if non-degradable or slowly degradable nanoparticles accumulate in organs, another concern is their potential interaction with biological processes inside the body: because of their large surface, nanoparticles on exposure to tissue and fluids will immediately adsorb onto their surface some of the macromolecules they encounter.

The large number of variables influencing toxicity means that it is difficult to generalise about health risks associated with exposure to nanomaterials – each new nanomaterial must be assessed individually and all material properties must be taken into account.

Health and environmental issues combine in the workplace of companies engaged in producing or using nanomaterials and in the laboratories engaged in nanoscience and nanotechnology research. It is safe to say that current workplace exposure standards for dusts cannot be applied directly to nanoparticle dusts.

Environmental Issues

Nanopollution is a generic name for all waste generated by nanodevices or during the nanomaterials manufacturing process. This kind of waste may be very dangerous because of its size. It can float in the air and might easily penetrate animal and plant cells causing unknown effects. Most human-made nanoparticles do not appear in nature, so living organisms may not have appropriate means to deal with nanowaste.

Fig. Groups opposing the installation of nanotechnology laboratories in Grenoble, France, have spraypainted their opposition on a former fortress above the city

To properly assess the health hazards of engineered nanoparticles the whole life cycle of these particles needs to be evaluated, including their fabrication, storage and distribution, application and potential abuse, and disposal. The impact on humans or the environment may vary at different stages of the life cycle. Environmental assessment is justified as nanoparticles present novel (new) environmental impacts. Scrinis raises concerns about nano-pollution, and argues that it is not currently possible to "precisely predict or control the ecological impacts of the release of these nano-products into the environment."

On the other hand, some possible future applications of nanotechnology have the potential to benefit the environment. Nanofiltration, based on the use of membranes with extremely small pores smaller than 10 nm (perhaps composed of nanotubes) are suitable for a mechanical filtration for the removal of ions or the separation of different fluids. Furthermore, magnetic nanoparticles offer an effective and reliable method to remove heavy metal contaminants from waste water. Using nanoscale particles increases the efficiency to absorb the contaminants and is comparatively inexpensive compared to traditional precipitation and filtration methods.

Furthermore, nanotechnology could potentially have a great impact on clean energy production. Research is underway to use nanomaterials for purposes including more efficient solar cells, practical fuel cells, and environmentally friendly batteries.

A Need for Regulation?

Significant debate exists relating to the question of whether nanotechnology or nanotechnology-based products merit special government regulation. This debate is related to the circumstances in which it is necessary and appropriate to assess new substances prior to their release into the market, community and environment.

Regulatory bodies such as the United States Environmental Protection Agency and the Food and Drug Administration in the U.S. or the Health and Consumer Protection Directorate of the European Commission have started dealing with the potential risks posed by nanoparticles. So far, neither engineered nanoparticles nor the products and materials that contain them are subject to any special regulation regarding production, handling or labelling. The Material Safety Data Sheet that must be issued for some materials often does not differentiate between bulk and nanoscale size of the material in question and even when it does these MSDS are advisory only.

Limited nanotechnology labeling and regulation may exacerbate potential human and environmental health and safety issues associated with nanotechnology. It has been argued that the development of comprehensive regulation of nanotechnology will be vital to ensure that the potential risks associated with the research and commercial application of nanotechnology do not overshadow its potential benefits. Regulation may also be required to meet community expectations about responsible development of nanotechnology, as well as ensuring that public interests are included in shaping the development of nanotechnology.

HEALTH AND ECOLOGY

In a witty review of popular attitudes towards food, David Cort remarks that he has encountered absurdly conflicting dietary recommendations in books and articles on nutrition. One authority advises his readers to avoid fat; another advises that it should be eaten, on the theory that fat is the least fattening of all foods. Potatoes have been alternately damned and praised; mixed meals have been both frowned upon and approved; the consumption of alcohol, criticized and recommended. "The cruelty and irresponsibility of offering one single, standard dietary solution for everybody is obvious," Cort adds.

"What a person eats is, in many important respects, his life. But every individual is different from every other individual. No doctrinaire solution will work for them all. Each individual has conditioned habits of eating, of taste, of appetite, of expenditure of energy, of nervous rhythms and, most important of all, of metabolism. Few modern doctors have the interest, time or genius to find out all about any one individual; and many are themselves overweight. But the individual has the time and interest and, at least about himself, perhaps the genius. After all, it's his life."

Although Cort's remarks are confined to nutrition, they are quite relevant to a number of problems in human ecology. It would be an error to form rigid concepts of a normal man, a normal diet, or a normal way of life. Any such image would be woefully lacking in biochemical and physiological support. The biologist and physician must still admit, as Carrel did nearly a quarter of a century ago, that "most of the questions put to themselves by those who study human beings remain without answer. Immense regions of our inner world are still unknown." Impressive advances, to be sure, have been made during the past two decades in solving major problems of cellular biochemistry, endocrinology and physiology. Knowledge of the transformation of energy in the cell is very far advanced, for example and great strides have been made in understanding the chemistry of the genetic material in the cell's nucleus. Despite this newly acquired knowledge, however, nutritionists have an extremely limited understanding of the function of many trace elements in the body and of the way in which vitamins prevent deficiency diseases. The nutritional requirements of the various tissues, essential details of the blood-forming system, the mechanisms of resistance to disease and many key relationships within the body await clearer understanding.

The human organism is extremely sensitive to very small quantities of hormones and nutrients. A variation by only a few drops in the daily output of insulin, a pancreatic hormone, will determine whether an individual will handle sugar properly or succumb to diabetes. A difference of a few thousandths of a gram (milligrams) in the intake of certain vitamins and trace elements makes for either well-being or nutritional disorders. According to the National Research Council, a man twentyfive years of age requires a daily intake of 1.6 milligrams of thiamine and riboflavin, 7.5 milligrams of vitamin C and 12 milligrams of iron to preserve good health.

His minimum daily requirements of iodine and copper are believed to be as little as 0.1 and 0.2 milligrams, respectively. If the daily intake of these nutrients is reduced by a few thousandths of a gram or less for a long period of time, the result will be a marked impairment of health. Despite the minute quantities involved, deficiencies of vitamins and minerals occur on a fairly large scale in the United States. If A. F. Morgan and L. M. Odland are correct in their claim that a dry, roughened skin and lesions of the eyes and mouth "may be considered a useful pointer to possible dietary faults," about 50 per cent of the adolescents examined in the affluent northeastern states and 20 per cent in the western states may have nutritional disorders.

We tend to look upon diseases as though they were localized disturbances, confined to definite parts of the body. Although it is customary to acknowledge the body's unity in sickness and in health, the whole is often overlooked in favour of the parts and vital interrelationships are ignored in favour of isolated events. Actually, every major deficiency and physical impairment has far-reaching effects. "Disease consists of a functional and structural disorder," Carrel notes. "Its aspects are as numerous as our organic activities. There are diseases of the stomach, of the heart, of the nervous system, etc. But in illness the body preserves the same unity as in health. It is sick as a whole. No disturbance remains strictly confined to a single organ. Physicians have been led to consider each disease as a specialty by the old anatomical conception of the human being. Only those who know man both in his parts and in his entirety, simultaneously under his anatomical, physiological and mental aspects, are capable of understanding him when he is sick."

A clear understanding of illness and health requires a greater appreciation not only of man but of men. Physical needs and responses vary enormously from one individual to another. Roger J. Williams, one of America's most eminent biochemists, emphasizes that within the same racial groups, even in the same families, men differ from one another in the shape and weight of their vital organs, the composition of body fluids, the output of endocrine glands, the rate of metabolism, temperature control and many other respects.

Individual differences in the amount of nutrients required for health and in the amount of nutrients absorbed by the body are of major practical importance. Individual requirements of calcium, trace elements, amino acids and vitamins are highly variable. For example, two normal five-year-old children who ate the same food and were exposed to the same environmental conditions showed a great variation in the amount of calcium they retained from their diets.

One child retained 264 milligrams of calcium per day for 45 days, whereas the other retained 469 milligrams, a difference of nearly 80 per cent. A high degree of variability is also found in individual responses to common environmental toxicants. Williams cites an account of 78 men who were exposed to vapors of carbon tetrachloride, of whom 15 suffered acute poisoning. Six became sufficiently ill to require hospitalization. "All were white men within 5 to 8 years of the same age and in general good health," observes Fredrich H. Harris, the source for Williams's account. "All 15 of the men poisoned were exposed from 3 to 8 hours; however, many of those who developed no symptoms were exposed for a similar length of time or longer."

The fact that man's knowledge of important bodily processes is incomplete and that many individuals exhibit marked deviations from "normal" responses and needs does not imply that the role of medicine is inconsequential in promoting health or that scientific generalizations are useless. The healthiest of men will undoubtedly require medical attention at some time in their lives and every physician must use physiological and anatomical generalizations as points of departure for the diagnosis of a disorder. Basic research in biochemistry and physiology promises to increase vastly our understanding of the preconditions for human health and well-being. But a phenomenon does not cease to exist because it cannot be readily explained by current methods and theories.

On the contrary, we are often profoundly affected by phenomena that are too complex to lend themselves to precise analysis at the present stage of scientific development. The way in which stress produces illness and the role that nutrition plays in the occurrence and prevention of chronic diseases constitute fields of research that have scarcely been penetrated. It would be a major setback if the roles of these factors were minimized simply because they present formidable problems to research. Man must try to conserve many practices that, on the basis of long experience, are known to promote good health, even if the reasons why they are beneficial cannot be stated in precise biochemical terms.

What are some of these practices? As it is clear that our bodily functions are interdependent, we should not permit any part of our biological equipment to atrophy. This means that we must use all the parts of our bodies. Our musculature must be as fully employed as our minds. Many activities and experiences that have been restricted by our sheltered civilization seem to play important roles in preserving health. There is a suspicion that modern man indirectly harms his body by failing to use certain physiological resources that once supported his ancestors during periods of hunger and cold.

"There exists in the body of man, as of all animals, biological mechanisms for the storage of food developed for meeting the irregularities and cyclical changes in nature," Rene Dubos writes. "It may still turn out that a nutritional way of living permitting continuous growth at a maximum rate may have unfortunate distant results. Fasting fads may have some justification after all by providing an opportunity for the operation of certain emergency mechanisms built by nature into the human body."

Although it remains to be seen whether this conclusion will be accepted by nutritionists and physicians, it is reasonably clear that a one-sided manner of life is biologically undesirable. An existence anchored in either muscular exertion or inactivity, surfeit or poverty, stress or an overly sheltered manner of life, tends to produce physiological disequilibrium. A prolonged period of muscular exertion causes marked damage, whereas protracted physical inactivity appears to increase both the effect and the incidence of heart and circulatory disorders. When the coronary arteries of dogs are surgically narrowed to simulate atherosclerosis, a lack of exercise makes it difficult for new blood vessels to form and provide more blood to nourish the diseased area of the heart muscle. There also seems to be more coronary illness among men whose occupations are sedentary than among those who engage in physical work. In contrast to the well-exercised individual, the "physically inactive individual shows signs of aging earlier in life," emphasizes Hans Kraus, of New York University. "He exists physiologically at a lower potential and is less well-equipped to maintain homeostasis [stable bodily conditions] and to meet daily stresses. This low level of function, combined with enforced suppression of the 'fight and flight' response, enhances the incidence of disease."

Not only does health require all the elements of a full life; it also requires that they exist in balance. A moderate amount of exertion, a balanced intake of nutriment and a reasonable amount of exposure to stress are necessary for the maintenance of well-being.

Almost any kind of excess is harmful. This has become apparent in respect to nutrition. "In laboratory experiments rats fed an unlimited diet were found to die sooner than animals prevented from gaining weight by a diet severely restricted in quantity but well balanced in composition," Dubos observes. "Likewise insurance statisticians have repeatedly emphasized that in man the obese have a short expectancy of life. Indeed, obesity is now publicized as the most common nutritional disease of Western society. This was apparently true also in Imperial Rome, as during all periods of great material prosperity.

'In the old days,' wrote Lucretius in the fifth book of De Rerum Natura, 'lack of food gave languishing limbs to Lethe; contrariwise today surfeit of things stifles us.' Thus, history repeats itself. Like the prosperous Romans of two thousand years ago, countless men of the Western world today are digging their own graves through over-eating."

Finally, if man is to maintain good health, he requires diversity. It is hardly necessary to emphasize that a healthy psyche and a rounded response to stress are nourished by a variety of experiences and by diversified surroundings. However, it does need to be emphasized that variety is also necessary for the satisfaction of man's nutritional needs. By limiting his selection of foods, he may fail to acquire nutrients known to exist in natural foods but not yet identified by chemists or nutritionists. "Because we know that unidentified factors exist in foods of plant and animal origin," observes George M. Briggs, of the National Institutes of Health, "it is wisest to eat a wide variety of foods from the many excellent food groups."

If this is done regularly, Briggs emphasizes, it is unnecessary to supplement the diet with vitamins and so-called health foods. It might be added that if food such as grain were not overly processed, it would be unnecessary to "enrich" bread with synthetic vitamins, which replace only part of the valuable nutrients lost in the milling process. In general, the more a food is processed, the greater the likelihood that identified nutrients as well as unidentified ones will be removed. Completeness, balance and diversity should be regarded as practical ecological concepts—as important in producing healthy human communities as they are in producing stable plant-animal communities. Indeed, it is not farfetched to say that when these concepts are correctly applied, they promote human health because they produce a stable ecosystem of men, animals and plants. In the last analysis, the ecology of health is grounded in natural ecology. A complete way of life for man presupposes unrestricted access to the countryside as well as the town, to soil as well as to pavement, to flora and fauna as well as to libraries and theaters. A balanced way of life presupposes a lasting equilibrium between land and city; animals, men and plants; air, water and industry.

Diversity presupposes an awareness that nearly every species perpetuates the stability of the biosphere, either directly or indirectly. As the city encroaches on the land, however, human life becomes increasingly restricted to a polluted, nerve-racking urban environment. As the city begins to dominate the land, physical activity yields to sedentary forms of work. Finally, as the aggregations of population swell to massive proportions, foods become laden with chemicals and human activities are standardized. By oversimplifying the natural environment, we have created an incomplete man who lives an unbalanced life in a standardized world. Such a man is ill—not only morally and psychologically, but physically.

ECOLOGY AND HUMAN BEHAVIOUR

Disease is an inevitable part of life and coping with disease is a universal aspect of the human experience. All humans, during the course of their lives, harbour infections by disease organisms and suffer the consequences of those infections. The experience of disease, by individuals or whole populations, is as inescapable as death itself. Yet the particular diseases that afflict people, as well as the way in which symptoms are interpreted and acted upon, vary greatly by culture. Understanding the nature of interactions between disease and culture can be a productive way of understanding humanity and is therefore an important topic in medical anthropology. Because of its biocultural, evolutionary and crosscultural perspectives, anthropology has much to offer to the understanding of the causes and consequences of disease. From an anthropological perspective, diseases cannot be explained as purely "things in themselves"; they must be analyzed and understood within a human context—that is, in relation to ecology and culture.

The distribution of disease in a population is neither constant nor random. The diseases that characteristically afflict members of a population vary significantly among societies because of differences in culture, ecological setting and historical period. More important, within a single society, there may be striking variations in the kind and severity of diseases that afflict individuals of different ages, sexes, social classes and ethnic groups. Understanding the descriptive epidemiological distribution of disease morbidity and mortality is important to medical anthropologists particularly because the social patterning of disease distribution often reflects the cultural coding of behaviours. Understanding the role of culture in disease distribution is also necessary for the implementation of successful disease control programmes

Culture plays a major role in determining the patterns of disease and death in a population for two reasons. First, culture may shape important behaviours (with respect to diet, activity patterns, water use, sexual practices, etc.) that predispose individuals to acquire certain diseases. Second, through culture, people actively change the nature of their environment, often in ways that affect their health. The archaeological and historical record clearly demonstrates that the environmental changes caused by humans can have profound effects, both positive and negative, on disease rates. Although humans have a dual system of inheritance through both genes and culture, culture is the primary mechanism for survival. Culture is a mechanism of adaptation to environmental threats, such as diseases, which act as agents of natural selection in the evolution of both human biology and culture. Cultural practices, however, can also be maladaptive when they exacerbate health problems.

Ecology is the study of the relationship of organisms in an environment. Human societies coinhabit their environment with many other organisms, including those producing disease. An ecological approach to human health and illness emphasizes the fact that the environment and its health risks are, to a significant extent, created by the culture. In many cultures, people think of themselves as masters of their environment, because they exploit so many plants and animals within the food chain as sources of energy and nutrients. Yet at the same time, humans are being exploited by microorganisms, including those that cause disease, as a source of food and shelter. Disease ecology primarily focuses on the multiplex interactions of the pathogen, the environment and the human host. The nutritional, physiological, genetic and mental condition of the human individual host all play significant, if still incompletely understood, roles in both infectious and non-infectious disease states. The current international health problems called "emerging infections" provide a good example of the mutability of these ecologic interactions.

The study of disease and human behaviour in an ecological setting is a fundamental task for medical anthropology. The approach contributes to basic and applied research in the field by providing a strategy for answering some of the major questions raised by both general anthropology and epidemiology. For example, it can be applied to anthropological questions concerning the interaction of biology and culture in human evolution or to questions of why particular cultural behaviours may "make sense" and be retained in an ecological setting.

This type of research strategy is truly biocultural and can therefore help to bridge the gap between biological and cultural anthropology. In epidemiology, the contribution of an anthropological focus on human behavioural patterns can help unravel fundamental questions of disease causality. The study of disease, ecology and behaviour also has important implications for public health programmes. Through the study of behavioural patterns related to the social epidemiology of disease, it is possible to design health programmes that are both effective and culturally acceptable. Moreover, the ecological approach can help to anticipate the health implications of technological change or new political-economic policies.

This chapter discusses basic concepts and methods in the study of disease, ecology and behaviour; it is not an exhaustive review of the literature. Instead, we summarize many illustrative examples of research that focuses on particular diseases in particular cultural settings. These examples emphasize infectious diseases for several reasons: first, the etiology (causation) of these diseases is the best understood; second, the ecological approach described here is most applicable to this category of disease, (although it is also applicable to the study of chronic diseases with complex etiologies, which are characteristic of industrialized societies, as well as nutritional deficiency diseases, which are characteristic of the poor in economically developing societies); and, finally, infectious diseases still represent the major cause of morbidity and mortality in non-industrialized societies, which anthropologists have traditionally studied.

DISEASE AND HUMAN BEHAVIOURAL PATTERNS

As the field of epidemiology has made clear since its inception in the late 1800s, diseases are not distributed randomly in human populations. Some individuals—and some groups of individuals—are at increased risk from various diseases, for masons that are often unclear. Epidemiologists not only describe patterns of disease occurrence through space and through time, but they attempt to elucidate disease etiology through the search for risk factors that appear to be significantly associated with disease outcome. Disease risk factors are of two major types. Endogenous risk factors are those that are biologically intrinsic to the human host. For example, genetic diseases, such as sickle cell anemia or hemophilia, have, by definition, an endogenous etiology. More commonly, however, genetic inheritance implies a predisposition to a disease that requires other variables or cofactors for expression to occur.

Exogenous risk factors are those that am extrinsic to the body of the human host. Some of these may be biotic, such as microorganisms that cause infectious diseases; others are non-biotic substances present in the environment, such as toxic chemicals in the workplace. In most cases of disease, both endogenous and exogenous factors are involved—hence, the notion of "multiple causation," or "multifactorial etiology". Humans may unwittingly increase the likelihood of disease by exposing themselves or others to risk factors of both the exogenous and endogenous variety. In many cases, this enhanced exposure potential occurs through disruption of existing ecological relationships between the host, the agent(s) of disease and the environment. In this way, human behaviour itself may be said to be a risk factor for disease, in that human activity may be a necessary component in the chain of events leading to a disease outcome.

Anthropologists, as professional observers and interpreters of human behaviour, have an obvious and crucial role to play in the understanding of disease etiology: they can facilitate risk factor identification by describing distinctive patterns of human behaviour related to the social distribution of disease. In this capacity, anthropologists may contribute directly to the generation of causal hypotheses, as they did in the case of kuru and cannibalism in New Guinea. In addition, anthropological descriptions of risk factor exposure based on long-term ethnographic observation may be more valid than those normally obtained through the standard epidemiological technique of questionnaire surveys.

Perhaps most important, anthropologists are especially equipped to understand disease-promoting human behaviours in sociocultural context. This includes the distribution of these behaviours through space and time, as well as the ideological and political-economic factors that serve to legitimate these behaviours. It is in this latter capacity—as interpreters of human behaviour who elucidate how and why people act the way they do—that anthropologists may contribute directly to medical anthropological theory building and indirectly to disease prevention and control.

PUBLIC ROLE IN HEALTH CARE MATTERS

Health care in about 1990 cost at least $1.7 trillion, or about 8 per cent of world income, making it one of the largest industries in the global economy. On average, 60 per cent of this is public spending. If this spending is excessive or otherwise inappropriate, the consequences for the economy and for health outcomes could be substantial. Governments also provide a large share of health services, sometimes as large as the share in spending, and often intervene in various ways in the private health care market. Since most health care is a private good, it is surprising that so much of it is provided, financed or regulated by the state. In contrast to what happens in many other sectors of the economy, this substantial public role is most pronounced in high income countries which are generally very market-oriented; the state usually finances a smaller share of health care in poorer countries.

Does this pattern provide models for the future development of the health sector and the public role in it, for lowincome countries? And is the variation among countries in the amount and kind of public intervention associated with differences in people's health, in what is spent on health care or in how well health systems function?

Besides consuming large resources, many health systems are regarded as inefficient or inequitable or both; they are often described as in "crisis", as needing "reform", or as having "failed". Are the supposed failures of health systems real? If so, they might be caused by misguided public intervention, so they could be corrected by a smaller or different public role and a greater reliance on private markets. Or governments might intervene for sound reasons, to correct or compensate for failings in those markets; that is, outcomes would be even worse if left entirely to the private sector. Is there an appropriate frontier between private and public action, and a best combination of instruments for the state to use when it intervenes? These and related questions must be confronted in any reform effort. It is relatively simple to conclude that governments should do certain things and should leave others to private activity; but often there is a variety of possible solutions and no obviously best approach. Theory does not always provide clear answers, and the empirical evidence is incomplete, extremely varied and difficult to interpret.

CHOICES FOR STATE INTERVENTION

It matters not only *whether* governments intervene, but also *how* they do it: the second essential question is what the public sector should do, given that some problem in the private market appears to warrant some public action. This is particularly important because government failings in the health sector are also common, and often result from intervening in the wrong ways or with the wrong instruments.

There are five distinct instruments of public intervention: arranged from the least to the greatest intrusion into private decisions, these are to:

1. *Inform,* which may mean to persuade, but does not require anyone to do anything. Governments do this when they publicize the health risks of smoking, or include health and basic hygiene education in public schools. These are examples of information directed at consumers, but governments also inform health care providers and suppliers of health care inputs, as by conducting research and disseminating information on disease patterns and on the effects and risks of medical procedures.
2. *Regulate,* which determines how a private activity may be undertaken. Governments sometimes regulate the medical profession by setting standards for doctors or accrediting hospitals, although these activities may also be undertaken by private bodies. And government regulation is common in the insurance industry, in the importation of medical equipment, drugs and supplies and in the protection of food and water quality. More generally, governments can influence private health care activity in many ways, often combining regulation with some financial incentives to offset the costs but without public financing. Regulation is usually pursuant to a law, and is often determined by an executive or administrative body.
3. *Mandate,* which obligates someone to do something and to pay for it. Compliance with regulations can also imply substantial private costs; but a mandated activity is different in that it *must* be performed, whereas a private producer can react to regulation by choosing not to undertake the activity. Mandates are usually specified in law, which may subsequently be adumbrated by regulation. The most important mandates, in financial terms, are the requirements that employers provide health services or insurance to their employees, or contribute to social insurance funds for that purpose. Governments can also impose mandates on individuals, as by requiring that children entering school be immunized.

4. *Finance* health care with public funds. Because mandated insurance is effectively paid for by an earmarked, involuntary contribution which is equivalent to a tax, "public" health expenditure is commonly defined to include such costs along with expenditures from public budgets. The obverse of spending public funds is to tax particular activities or goods, such as alcohol and tobacco, at least partly for health reasons. This issue is not treated here; while taxation which reduces consumption of specific goods may have substantial health effects, it is always limited to very few goods and is not used systematically to promote health. Finally, the state may—
5. *Provide* or deliver services, using publicly-owned facilities and civil service staff. This is what Ministries of Health in most poor countries do; so do various governmental bodies in many countries at all income levels. Once a society has decided to finance health services with public funds, the choice arises of whether to provide them through public facilities or to pay private producers to provide them. The appropriate way to consider this choice is as a standard "make or buy" decision. The issues for a government are the same as for a private firm, and turn on costs—is it cheaper to produce something than to buy from an outside supplier?—and on the risks and difficulties of enforcing contracts and avoiding fraud when dealing with such suppliers.

Because public financing requires public resources, it is perhaps the crucial choice about state action. However, all the instruments mentioned have costs; even information is not free. The benefits from any intervention have always to be weighed against these costs.

In addition, sometimes two instruments overlap, or one requires the use of another: mandates imply regulation, and public provision usually implies at least partial public finance. And two instruments can be alternatives: for part of the population, governments can either finance health care or mandate financing by employers or other private institutions. Differences in ability to pay make it natural to operate mixed systems of public intervention, such as mandated coverage for the non-poor and public finance for the poor. There are alternatives even within mandated coverage, such as "play or pay" arrangements in which employers can finance health care directly for their employees or pay into a social security scheme.

THE APPROPRIATE PUBLIC ROLE IN HEALTH

As a first approximation, it is easier to say what govern-ments *should not do* in health than to specify what they *should do.* That is, it is clear that certain actions are likely or certain to violate one or more of the objectives of a health care system to an important degree. To apply a topographic metaphor: such actions correspond to falling off a plateau of satisfactory outcomes, all somewhat different but none clearly dominating the rest, and into one of the surrounding chasms of one or another kind of failure. Four "don'ts" are discussed in what follows: they refer to the way the public is taxed, charged or exempted to pay for health care; the providers to which governments transfer public funds; the way providers are paid; and the services they are paid to provide.

WHAT GOVERNMENT SHOULD NOT DO

The first thing governments should not do is to use the tax system, or any system of fees at public facilities, to make the poor subsidize the health care of the rich. Conceptually, subsidies are justified only for the poor, and broader financing of care through insurance is a question of efficiency rather than of equity. This is not only a matter of whether the rich use more of publicly-financed services than the poor do, although great inequities often arise because the rich have more access to those services. Financial equity also depends on who pays the taxes. More narrowly, governments should not contribute to social security financing from general revenues, unless coverage is universal, because when only part of the population is covered it is usually the poor who are excluded. And governments should not treat private insurance coverage as a cost to employers unless it is also treated as income to beneficiaries. Such practices are not only inequitable; they are also inefficient to the extent that they lead to excessive spending on health care, or reduce labour mobility.

Controlling inequitable subsidies does not mean that social security systems should be dismantled or must be made universal. Even when incomplete, such mandated insurance includes substantial progressive subsidies from high-paid to low-paid workers, and the mere fact that some people in society receive more generous health care is not necessarily a problem, so long as they pay for it. What matters is that governments not make everyone pay for what some are excluded *ex ante* from receiving. Perverse subsidies not only cause immediate inefficiencies or inequities, they also create interests that oppose subsequent health system reform. This is evident in the United States and Chile and it is the reason why the design of subsidies is a crucial part of any reform to extend or improve coverage for the poor. The second thing governments should not do is tie public finance to public provision. The choice of whether to provide care through public or parastatal facilities should be treated as a "make or buy" decision, subordinate to the larger decision about what to pay for. That does not necessarily mean eliminating public provision, which will sometimes be the best solution. It means rather that competition between public and private providers should be based on costs and on quality, and not on price to the consumer, as is commonly the case.

To achieve those goals, however, requires other changes in how public institutions operate, changes that probably cannot occur so long as public funds go automatically and exclusively to government facilities. This conclusion is most pertinent to poor countries where public systems are most likely to be inefficient and to be used only because they are free or nearly so. High-income countries with a large share of public provision, particularly in hospitals, suffer less from these problems. A third thing governments should not do is pay for health care by fee-for-service, unless other mechanisms are used to control expenditures. This is seldom a problem with publicly provided services, but makes it hard to control costs when governments finance private providers. Even negotiating or controlling the fees is not enough, as the Canadian experience demonstrates, since providers can respond to fees they consider too low, by increasing output. This helps explain why Canadian health expenditure has risen faster than that in European countries which rely on other payment mechanisms.

European countries which pay for health care by fee-for-service also rely on global budgets, utilization reviews or other instruments of cost containment. Since expenditures equal prices times quantities, and since quantities of services respond to prices and are difficult to control directly, countries which pursue both macroeconomic and microeconomic efficiency in health spending usually control both prices and expenditures. In contrast to European and North American experience, Brazil applies this system to block grants for federal expenditure in states with relative financial autonomy; but in states where federal money is paid directly to providers, the federal government also controls one crucial quantity, the number of hospitalizations that federal money will pay for. In all cases, the government sets prices for services.

It also helps, in controlling expenditures, to define "services" as complete treatments for specific conditions rather than as all the individual components of such treatments, so they can be financed by mechanisms such as the Diagnostic Related Group payments used in the United States and the provider can be required to assume part of the financial risk. Changing from a system based on overall budgets and on salaries for providers, to a fee-for-service system without offsetting the resulting incentives to overprovision can lead to an explosion of costs, as in the Czech Republic. Finally, if governments mean to pursue some combination of better health and lower costs, they should not—in fact they cannot—simply finance whatever people demand when care is free to consumers.

This does not mean governments should not "subsidize demand" rather than "subsidizing supply" or providing services. It means only that there must be limitations on what will be paid for publicly. Such limitations are a particularly contentious matter, because both good health and cost containment may be opposed to consumer or provider satisfaction, which are also politically important objectives.

Nonetheless, two empirical observations are germane to this decision. One is that private insurance always carries some limitations, either as to the services covered or as to cost-sharing. Except where poverty is important, so that cost-sharing is more difficult, there is no reason for public finance to be systematically more generous in this regard than private risk-sharing arrangements. The second observation is that private markets will tend to supply what people demand, and public intervention typically acts to emphasize needs instead. In other forms of public subsidy, it is common to distinguish between wants and needs, and concentrate spending on the latter. This is the case, for example, with food subsidies, which are also very much healthrelated. Price subsidies are usually limited to "basic" foodstuffs, and food stamps cannot legally be used to buy alcohol or other non-necessities.

This may be considered partly a matter of efficiency, assuring more health gain than would otherwise occur. But it is also a matter of equity: in contrast to actuarial private insurance, where every purchaser buys the expected value of the health services needed, public finance is involuntary. It comes from taxpayers who have a legitimate interest in meeting needs, and thereby getting value for their money, but not necessarily in paying for wants.

None of these "don'ts" is easy to implement, because some consumer or provider interests can be expected to oppose every one of them, in every country. But they are arguably the most important conclusions about where the public/private frontier should run in the health sector. It is notable that none of these conclusions depends solely on features unique to the health sector. The peculiarities of health matter most for the difficulty of distinguishing "needs" from wants, and—partly as a consequence—for the dangers of paying providers for whatever they choose to provide, without incentives to control costs. A system which avoided the problems would still present complex and difficult questions for the proper role of the state, but it would have more latitude to pursue improvements. And it would not matter so much, exactly what objectives the government pursued or which combination of instruments it applied.

WHAT GOVERNMENT SHOULD DO

Beyond the prescriptions for how the state should deal with the problems of each of the three domains of health care, several "dos" appear to be generally valid for governments. If the objective is to minimize deadweight losses from public intervention and leave as much room as possible for private choices, then the first thing governments should do is to use each less-intrusive instrument to the point where a more intrusive intervention is justified, following the sequence of increasingly greater interference— inform, regulate, mandate, finance and provide services. That is, governments should regulate private activity when merely improving people's information is not enough, deliver services if it is infeasible to finance private providers equitably, and so on. Public finance is inescapable for some actions, but particularly in low- and middle-income countries much can probably be accomplished by better use of information and regulation. Failure to use these other instruments well, can increase the need for public finance. Sometimes the problem is that governments exploit these instruments too little. They do not regulate private insurance when it first begins to expand, which makes subsequent regulation politically more difficult; or they do not initially react when health-damaging behaviours such as smoking become more entrenched. Sometimes the problem is inappropriate regulation, which needlessly restricts competition, or enforces inefficiency in the public sector by centralizing nearly all decisions. And poor countries in particular often get the worst of both worlds by paying for activities which should be, but are not, regulated: government subsidy of medical education without adequate control of quality or relation to needs is a common example. In the worst of cases, governments use all the available instruments in exactly the reverse order. They try to provide more health care than they can pay for, with the result that most services are under-financed and of poor quality; they try to finance services, some of which might be mandated and paid for by consumers or employers; they mandate care, as by social security systems, without adequately regulating it; and they do too little to inform the public and providers either of dangers to health or of how the health care system is actually working.

Much of the criticism of government failure in the health sector, especially in poor countries, describes the result of getting things backwards. These ideas are represented in Figure, which shows the appropriate relation among the instruments by which the state can intervene in health care which portrays the kind of inappropriate or imbalanced relation often found in poor nations.

Figure indicates that whatever is mandated, financed or provided publicly should also be regulated, and much else besides. The state can finance some care that is not anyone's mandated responsibility, and can also mandate care or insurance coverage which is financed privately. And most if not all public provision should be fully financed by government, but possibly with much more scope for finance than for provision by the state.

IMPACT OF PESTICIDE ON HEALTH

A pesticide is any chemical which is used by man to control pests. The pests may be insects, plant diseases, fungi, weeds, nematodes, snails, slugs, etc. Therefore, insecticides, fungicides, herbicides, etc., are all types of pesticides. Some pesticides must only contact (touch) the pest to be deadly. Others must be swallowed to be effective. The way that each pesticide attacks a pest suggests the best way to apply it; to reach and expose all the pests. For example, a pesticide may be more effective and less costly as a bait, rather than as a surface spray.

TYPES OF PESTICIDES

BACTERICIDES

A bactericide or bacteriocide is a substance that kills bacteria and, preferably, nothing else. Bactericides are either disinfectants, antiseptics or antibiotics.

FUNGICIDES

Fungicides are chemical compounds used to prevent the spread of fungi or plants in gardens and crops, which can cause serious damage resulting in loss of yield and thus profit. Though oomycetes are not fungi, they use the same mechanisms to infect plants and therefore in phytopathology chemicals used to control oomycetes are also referred to as fungicides.

Fungicides are also used to fight fungal infections.Fungicides can either be contact or systemic. A contact fungicide kills fungi when sprayed on its surface; a systemic fungicide has to be absorbed by the plant.Fungicide residues have been found on food for human consumption, mostly from post-harvest treatments. Some fungicides are dangerous to human health, such as Vinclozolin, which has now been removed from use.Like other pesticides, fungicides can induce pesticide resistance. Equivalently, antifungal drugs can induce drug resistance.

HERBICIDES

A herbicide is used to kill unwanted plants. Selective herbicides kill specific targets while leaving the desired crop relatively unharmed. Some of these act by interfering with the growth of the weed and are often based on plant hormones. Herbicides used to clear waste ground are nonselective and kill all plant material with which they come into contact.

Some plants produce natural herbicides, such as the genus Juglans (walnuts). Herbicides are widely used in agriculture and in landscape turf management. They are applied in total vegetation control (TVC) programmes for maintenance of highways and railroads. Smaller quantities are used in forestry, pasture systems, and management of areas set aside as wildlife habitat.

INSECTICIDES

An insecticide is a pesticide used against insects in all developmental forms. They include ovicides and larvicides used against the eggs and larvae of insects. Insecticides are used in agriculture, medicine, industry and the household. The use of insecticides is believed to be one of the major factors behind the increase in agricultural productivity in the 20th century. Nearly all insecticides have the potential to significantly alter ecosystems; many are toxic to humans; and others are concentrated in the food chain. It is necessary to balance agricultural needs with environmental and health issues when using insecticides.

Bibliography

A G Brecket: *A Handbook on Nanotechnology*, Dominant Publication, Delhi, 2008

Abdullahi, A.: *Statistical Mechanical Modeling and Its Application to Nanosystems*, Philadelphia, PA: University of Pennsylvania Press, 1998.

Ach J.S. and Siep L.: *Nano-bio-ethics: Ethical Dimensions of Nanobiotechnology*, Transaction Publishers London, 2006.

Ahmad, Lone: *Nanoscience: Nanotechnologies and Nanophysics*, South Kashmir Press, Jammu Kashmir, 2006.

Ajay Kakkar: *Know All About Nanotechnology*, Abhishek Publications, Delhi, 2010.

Allen, Robin: *An Introduction to Micro and Nanotechnology*, Oxford University Press, New York, 2002.

Allhoff F. and Lin P.: *Nanotechnology and Society: Current and Emerging Ethical Issues*, Springer The Netherlands, 2009.

Anand Prakash Singh: *Nanotechnology : A Broad Spectrum*, Axis Publications, Delhi, 2010.

Anthony Turner: *Nanotechnology : An Introduction*, Abhishek, Delhi, 2008.

Ashish Bharadwaj: *The Nanoscope : Encyclopedia of Nanoscience and Nanotechnology*, Pentagon Press, Delhi, 2005.

Awasthi, K P: *Advanced Magnetic Nanostructures*, Cyber Tech Publications, Delhi, 2008.

B K Parthasarathy: *Introduction to Nanotechnology*, Isha Books, Delhi, 2007.

Bainbridge, W.S.: *Societal Implications of Nanoscience and Nanotechnology*, UK: Intellect Books, Delhi, 2001.

Balakrishna Rao and K. Krishna Reddy: *Nanotechnology in Health and Life Sciences*, Campus Books International, Delhi, 2007.

Baldev Raj, B.S. Murty, P. Shankar, B.B. Rath and James Murday: *Textbook of Nanoscience and Nanotechnology*, Orient Blackswan, Delhi, 2011.

Basnet, J.: *Unique Features and Properties of Nanostructured Materials*, Indian J Community Med, Delhi, 2008.

Bharadwaj, Ashish : *The Nanoscope: Encyclopedia of Nanoscience and Nanotechnology*, Pentagon Press, Delhi, 2005.

Bhatnagar, Vasudev: *Nanostructure Science and Technology*, Campus Books, Delhi, 2009.

Bhatt, R.C.: *Nanostructure and Nanoculture*, Swastik Publication, Delhi, 2009.

Boucher P. M.: *Nanotechnology: Legal Aspects,* CRC Press Boca Raton, 2008.

Branda Paz: *A Handbook on Nanoelectronics*, Dominant, Delhi, 2008.

Brune: *Nanotechnology: Assessment and Perspectives*, Springer, Delhi, 2010.

Bueno, O.: *The Drexler-Smalley Debated on Nanotechnology: Incommensurability at Work*, Willey, Delhi, 2004.

Carl C. Koch: *Nanostructured Materials: Processing, Properties, and Potential Applications*, Noyes Publications/William Andrew Pub., Norwich, 2002.

Cleland, Andrew N.: *Foundations of Nanomechanics: From Solid-State Theory to Device Applications*, Springer, New York, 2003.

D.K. Sexena: *Nanotechnology Development,* Pearl Books, Delhi, 2012.

D.V.S.S.R. Prakash: *Encyclopaedia of Nanotechnology*, Anmol, Delhi, 2006.

Daniel Minoli: *Nanotechnology Applications to Telecommunications and Networking,* Wiley India, Delhi, 2011.

Datta, S., D.B. Janes: *The Handbook of Nanostructured Materials and Nanotechnology*, San Diego: Academic Press, Delhi, 1998.

David J.: *Micromachines and nanotechnology: the amazing new world of the ultrasmall*, N.J., Dillon Press, Parsippany, 1995.

Deura, Vijay: *Nanotechnological Applications in Cancer Treatment*, Global Vision, Delhi, 2008.

Devoret, M.H.: *Single Charge Tunneling in Coulomb Blockade Phenomena in Nanostructures*, Plenum Press, New York/London, 1992.

Diwan, Parag and Ashish Bharadwaj: *The Nanoscope: Encyclopedia of Nanoscience and Nanotechnology*, Pentagon Press, Delhi, 2005.

Don Norman: *A Student Handbook of Nanobiotechnology*, Dominant, Delhi, 2008.

Drexler, K. Eric.: *Engines of Creaton.* Anchor Press/Doubleday, Garden City, 1986.

Fahrner: *Nanotechnology and Nanoelectronics: Materials, Devices, Measurement Techniques,* Springer, Delhi, 2006.

Index